"There is no more thoughtful, observant, insightful, uplifting, and elegant writer in the English language than Stephanie Mills. Whether she is speaking of intricacies in nature, her personal travels and experiences, on the difficulties of the human/political condition, she lets the world in whole, brings forth meanings that are always unique, keen, deft, and often very funny. She is a classicist as a writer, an activist in spirit, ready to confront the juggernaut with considerable ferocity. *Tough Little Beauties* rewards the reader with insight into the difficulties we face as well as joy at the glory of it all."
—Jerry Mander, author, *In the Absence of the Sacred*

"*Tough Little Beauties* is a compendium of brilliant work spanning over twenty years. The collection contains essays not seen before in book form and reprinted work, including a fascinating journal of journeying in India and a healthy excerpt from the wonderful book *In Praise of Nature*. Whether readers are familiar or not with Stephanie Mills' writing they'll be happy to discover (or rediscover) a sizzling writer, firm in her beliefs, yet ceaselessly alert and questioning. The moral core of this book is thoughtful and solid, while its subjects take an impressive run through the likes of the relationship of humans to nature, techno-fantasy, parenting vs. not parenting, fatality and fetality, apocalypse, herpes, and Mother Teresa."
—John Keeble, author, *Nocturnal America* & *Yellowfish*

TOUGH LITTLE
BEAUTIES

STEPHANIE MILLS

Also By Stephanie Mills

Whatever Happened To Ecology (Sierra Club)
In Praise of Nature (Island Press)
In Service of the Wild (Beacon Press)
Turning Away From Technology (Sierra Club)
Epicurean Simplicity (Island Press)

TOUGH LITTLE BEAUTIES

Selected Essays

And Other Writings

STEPHANIE MILLS

Ice Cube Press
North Liberty, Iowa

Tough Little Beauties: Selected Essays and Other Writings

ISBN 9781888160307 (1-888160-30-6)

Library of Congress Control Number: 2007924514

Ice Cube Press (est. 1993)
205 North Front Street
North Liberty, Iowa 52317-9302
www.icecubepress.com
steve@icecubepress.com

Manufactured in the United States of America.

The paper used in this publication meets the minimum requirements of the American National Standard for Information Sciences—Permanence of Paper for Printed Library Materials, ANSI Z39.48-1992

The author gratefully acknowledges permission to reprint extended excerpts from *In Praise of Nature* edited and with essays by Stephanie Mills. Copyright 1990 by Island Press. Reprinted by permission of Island Press.

Author photo (cover and pg. 189) compliments of Gary Howe.

Cover art © 2007 Laura Waldo-Semken.

Photographs used for "In Her Woods" are by jeni l. spamer, an amateur photographer, cook, writer, archivist, and fan of Stephanie Mills. She can be reached at jeni.spamer@gmail.com

To Dad

"What would I recommend? Don't panic; appreciate your own lives; and try to help a little loveliness continue .... we're in for some pretty rough bumping over the next century. The important thing is that we don't totally lose our bearings and become degraded by it." —Sterling Bunnell

Contents

INTRODUCTION

As I write this, I'm reclining on the daybed in my writing studio. Out one window I see that a red pine sapling I planted is now a tree overtopping the roof. A reclusive bird sings a wheedling query. He's been active around here lately. Whatever his species, he's another regular, like the thrasher and the jay and the mourning doves. There's continuity in this little biotic community I inhabit.

The books shelved in here evidence the continuity in my intellectual community. They carry forward the work of colleagues, mentors, and intellectual idols. Their titles bespeak the subjects and the wisdom that matter: wildness, wilderness, and first peoples; evolution, extinction, and overpopulation; forest history, changes in the land, and ecological restoration; anarchism, nonviolence, voluntary simplicity, and bioregionalism.

My writing about these matters begins as conversation—often argument—within. Subjected to the discipline of prose, it may wind up as prayer. In my heart and in my work my equivocal humanism and unequivocal love of nature joust with my anger at our idiotic wastage of this ecosphere and squandering of our own promise within it.

Tough Little Beauties gathers writings grappling with that reality as well as others more personal. The title essay relates a visit to a rare population of dwarf lake irises on an island in Lake Michigan. Altogether, the pieces here span almost thirty years. They range about from Michigan to India, and address subjects that range from abortion and the Great Lakes fauna to the coming peak in oil and natural gas production.

I have lived in northwest lower Michigan since 1984. The writings in the section titled "The North Woods" detail and celebrate a sense of this place. The second section, "India Journals," describes journeys in a place radically different from my home. Although I had been abroad before, I hadn't left the West. For me, three weeks traveling in Jammu and Kashmir, Uttar Pradesh, and Tamil Nadu constituted an epoch.

The writings in "The Body Politic" rant about the politics of the human, particularly the female human body. It's not just the personal that's political. Along with cooperation and symbioses, there are intricate, unequal power relations pervading this life. I think, therefore I'm a feminist.

The essays under the heading "Religious Experience" were written during a valuable spell of theism. Although a divine apparatus no longer seems necessary to provide life with meaning or morality, I still have my religious pursuits. Chief among them is the "spiritual" swimming begun in San Francisco Bay and continuing in the lakes, great and small, hereabouts. "St. Herpes," also written at that time when I trusted my fate to a benevolent higher power, originated as a letter to my friend Felicia Guest. A leading sex educator, she asked me what it was like to carry that virus. Answering her question entailed some self-appraisal. I shared the results in hopes of alleviating the isolation and chagrin of my fellow vectors.

Turning from the personal to the planetary, "Earth and Her Discontents" treats of the catastrophic damage various civilizations have inflicted on the planet. It begins with some excerpts from essays that first appeared in the book *In Praise of Nature,* published in 1990 by Island Press. "Words for the Wild," a more recent essay, links extinction, globalization, and a critique of technology. "Fun While It Lasted," the latest piece in the collection, reports on Peak Oil, a resource crunch that bids to bring industrial civilization down. Cassandra-like, I've written to urge the public to take these problems seriously. The possibility remains that these severe crises could cause a breakthrough to a better way of being in the world.

In the sixties our elders blamed dissent on "outside agitators," as though a native critique of our country's conduct was inconceivable. The title of the concluding section, "Inside Agitation," puns on that. The first two of these essays were churned by inside agitation. The concluding piece, "Matrix of Solitude," arises from calmer, but no less engaged, introspection. The unexamined life may be worth living, but I haven't much experience with it. I would not care to be without my knowledge of self nor, for that matter, of history, ecology, and place. I am past middle age and beyond despair. I am a pessimist, yet I still do hope.

If the two most frightening creatures on the planet are mobs and dictators, there's an intermediate organism—the right-sized, self-managing, ecologically adapted community. Our species has long experience in forming these. They are a better way of being in the world than the one that has wrought such havoc. If we are to have a future in which there is reverence for life and respect for all beings, it will be in such communities. They will be shaped by people who were willing to change everything about their lives, worked like hell, and managed not to die too young. With any luck, they'll have forests and trees, diverse birds to wonder about, and even authors to carry forward the lore of how the people did it and what they learned.

—Completed under a blue moon in May of 2007
Maple City, Michigan

THANKS
Without Steve Semken's cordiality, good humor, and willing interest in publishing this collection, it would only be a faint gleam in its author's eye. Without Tom Dean's invitation to a conference at the University of Iowa, I'd never have met Steve. Without Tom's copyediting and Charlotte Robertson's senior editing, the book would be of lesser quality.

THE NORTH WOODS

THE JOURNEY HOME

(1997)

Some people have a talent for making themselves at home anywhere. The wanderers among us are like creatures of the air, naturally born to a life that isn't constrained by geography. More of us, I reckon, require familiar territory. Perhaps this kind of attachment to place is the product of humanity's long history of hunting and foraging, of tilling the soil, and of living or dying according to one's knowledge of place. For most of human experience, thriving has required a fine attention to the land's life.

Our rootedness has been steadily undermined by the advance of civilization, with the pace accelerating wildly during the fossil fuels episode. Whether they possess the desire or ability to settle in just any place, human beings are now set in motion about the planet like birds driven before a great wind. Some are seeking jobs, some "quality of life" in as-yet unspoiled environments. The frontier mentality, the hope that things will improve in the next territory, also works against rootedness. In a world beglamoured by mass media, Dallas or Beverly Hills may become a life's ambition. Place, we fancy, can be a matter of design.

In *The Media Lab,* Stewart Brand remarks that e-mail has destroyed the "tyranny of place"—an arresting phrase that contains a grain of truth. Like any great teacher, one's place can seem to be a tyrant. Until we slip the toils of flesh, we are ruled by our need to eat and breathe and drink and see, and for these we need homes with fertile soils, flowing waters, leaves trafficking in sunlight and air. We need nature's sound and motion to evoke our intelligence. These are real constraints, and the more our minds are entranced by getaway fantasies, fixated on big ideas, the more onerous the disciplines of place do seem.

Some claim that communities exist in computer networks. Hence these networks, devotees say, qualify as places, or "virtual" communities. Yet they entail few obligations and harbor no nesting birds or healing herbs. Their intelligence is spread very thin.

Genuine community requires that I deal with a mix of people, including some who at first blush are scary and incomprehensible, folks who don't seem to share any of my tastes, values, or concerns. What we do share is the fact of living in a certain place—Kasson Township, Leelanau County, Michigan, U.S.A.—and having to work out our differences, many of which have to do with land etiquette. Much as I would like to avoid such dealings—over hunting, dumping, off-road vehicle trespass, zucchini and tomato overproduction, runaway dogs, and found kittens—they are in my face and on my earth and I can't change channels or wad up the page to make them go away. Besides, the conviviality that geography imposes is good. In the ever-closer long run, as the global betrays the local and the local gets wise, neighbors will increasingly have to know and depend on one another.

For me, a place must also have soil and include life forms other than my own. I can't honestly regard my chosen intellectual cohort or even my own species as a full community. Place is habitat, the ecosystem that hosts the evolution of a whole association of organisms, from the minute and peculiar to the grand and cosmopolitan—from soil fungus to black bear, and thousands of species in between. Place is permeated and enlivened by flows of light, water, wind, and even seismic energy. Pulses of organisms—migrating birds, butterflies, fish returning to natal streams, mayflies, spring peepers, new generations of mice and voles, and the seasonal visits of hunters or gatherers—all swirl into the vital force of a place. In the country or in the city, the essence of place is wild.

In a world where commercial civilization deems land to be a commodity, governs human life, and mass-markets material culture, any refuge of diversity—topographic, genetic, or cultural—is worth protecting. Indeed, natural diversity should be held sacred and defended. But first, the particulars of place must be perceived. An exploration must be made.

If a place is more than a human thought, to have a sense of place requires the use of all one's senses, and sooner or later, muscle and bone. Semi-rural northwest lower Michigan, my home, demands and gratifies full sensory engagement. The weather here refuses to be ignored or minimized. I began this writing one late November day when it was snowing. The amazement of snow dawns afresh on me, a Californian transplant, every winter.

Snow's arrival begins to impose a white purity on the landscape and to supply the makings of a perfect calm. Early in the blanketing process, the skeletons of the summer's weeds are still exposed, their characteristic dark forms and different colors sketched against the white ground. The sooty, many-branched knapweed, its wiry stems set all akimbo, the comic pewter topiary of horsemint, the deep rusty pods of Saint John's-wort all seem brilliant amid the noncolor of cloudy skies and snow-covered earth. While I wallow in the minimalist beauty of this scraggly vegetation, part of what I now know about my place, and a clue to its history, is that most of the aforementioned are alien plants, profiting from disturbance in the land.

In place, one develops eyes to see and the strength to honor differences and detail. It takes both effort and epiphany. The everyday beauty that informs this beat up postglacial, de- and reforested land—the light streaming into the last golden leaves of a shrubby maple, the moist, lucid green of a patch of mosses, the moonlike stillness of a collapsed, abandoned anthill, the scintillation of a gossamer strand guying cherry twigs, and the sheer delicious clarity of the healing air—never fails to feast my soul, if I but partake.

Any place that is still hospitable to nature includes beings beyond number—from soil organisms churning out the humus to the mosses binding the sand to the chickadees preening the pines to the turtles, ants, squirrels, herbs and all the other flora and fauna. Around the county, the vagaries of human character are also rich. There's refuge and invitation in all that variety.

A sense of place requires some loose boundaries. No animal's home range is limitless or random. Ten years ago when I moved to the country I naively imagined that life would be easier than in the city. I was interested in *reinhabitation*, which, as defined by bioregionalist Peter Berg and biologist Raymond Dasmann, means "becoming fully alive in and with a place....

applying for membership in a biotic community and ceasing to be an exploiter." Thinking globally was making me crazy; for acting locally I felt I needed a snugger locale, a place I could comprehend. Small could be beautiful—and hopeful—because small can be observable, intimate, accountable. Small is also more readily exploitable and, alas, sometimes petty, although rarely as destructive as large.

Small communities have yet to attain the strength to escape the idolatry of growth. Power flows to the state, resources to the corporations, and local government can effect change only in the tiniest increments. Zoners must permit what are deemed reasonable uses of land. This creates big problems in rural locales like mine, where the uses of land determine our future, and so-called development forecloses a lot of possibilities. It's a widespread plight.

Suburban sprawl, roadbuilding, deforestation, replanting with alien trees, farming, grazing, and invasions of exotic plant and animal species may all be moving our ecosystem and its economy toward some simplified but precarious mean. As the natural world and the human cultural diversity it fosters are forced into a dull sameness, a wealth of local knowledge, skillful means, and decent subsistence is devalued and destroyed, and everyday living is reduced to getting and spending, hustling, or passivity.

Global civilization is no respecter of persons and contributes little to genuine identity. Where one is greatly influences *who* one is, for better or worse. Degraded places sap souls. Reclaiming and restoring some natural integrity is what we should be doing where we are. These activities are site-specific, take lifetimes.

It has become rare in the world for people to be born, live, and die in the same place where their ancestors' bones rest, in the earth that feeds and finally is fed by all. In the absence of some invitation or inclination to know and care, place may mean nothing more than where you wind up, may be just an address. Yet, rightly attended, place teaches us how to be human and reveals the saga of life and how to behave in organic circumstance. Loyalty to place—staying put—brings home the long cycles and grave truths of human life in intimate and inescapable ways, from new babies born to adult struggles for competence to elders needing care and passing on.

Witness enough weddings and funerals, graduations and divorces and you understand that you are being moved through time in an archetypal procession. In a steady context you see yourself and your neighbors learning and failing, suffering and flourishing, growing and dying. This slow experience contrasts the media focus on sports contests, celebrity peccadilloes, current events, and foreign affairs, none of which fosters compassion or maturation of the self. To be preoccupied with these elsewhere spectacles is to be homeless in the soul.

Residing in a real place, you must confront the changes in the land—patches of regeneration, swatches of erosion, woods reclaiming old fields, fancy homes palisading the shoreline. It takes some effort to learn to identify the members of the local forest. It takes years to see that the seasons are both consistent and varied and to learn to cope with the rigors they impose.

Every winter has its own character yet belongs to one great warrior tribe. A few years ago the winter solstice launched a full month of snows and temperatures below freezing. At the next solstice, warmth and light had long since melted away the teasing November snows. Last winter had bare ground and clear days throughout December. We dressed lightly and experienced a different winter, still clocked by the sun hung low on the horizon. Then snow came back that New Year's, demanding more layers of clothing, more wood for the stove, raising the question of what we will do for electricity when the coal runs out, what we will substitute for propane when oil production declines, and whether heating with wood is really a sustainable practice at our current level of population.

Reinhabitation is easier said than done, easier intended than effected. Those of us who have migrated to our locales from cities bring a thin wash of urban fantasy in our wake and inevitably affect the locale's character. We add our own splash of cultural confusion to the community, and bring momentum to its change.

Neither of the two little towns nearby consists of much more than a post office, a cafe, a grocery store, a gas station, and a tavern. Cedar and Maple City are innocent as yet of cappuccino and Thai cuisine and naughty lingerie shops, but it may just be a matter of time before these misplaced signs of urbanity and yupward mobility arrive. Traverse City, the region's economic hub 15 miles away, has gained all of the above in the past few years. There's not a slaughterhouse, a local rail system, or grange for miles around, though. Few locals under 70, myself included, eat a primarily seasonal diet, or know how to work a root cellar. Our food comes from supermarkets, we now think. The good news is the handful of organic farmers in the area and their commitment to rebuilding our soil, acre by acre. But the general dearth of basic knowledge marks an amazing decadence here in the solid, stolid, practical Midwest: quite an abrupt disconnection from earlier hardihood won providing for one's household from this land and in this climate, which set such strict terms. It is a radical dependency on far-flung lines of supply and grandiose, opportunistic systems of production and distribution.

That is not to say that the good old days served this place perfectly well. This county's old (circa 1850) European-American settlers brought some disastrously generic ideas about land use with them and the consequences are plain to see: what was a forest became farmland and woodlots and now is being encroached upon by strip malls and housing developments, a metastasis of practices inaugurated here in the name of civilization. Nevertheless, in their day the settlers understood the sheer necessity for homestead self-reliance and neighborliness. A good many of them were, till mid-century, able to maintain those values. They figured out just enough about the land to make a subsistence living—subsistence, but not luxury or leisure. Even at its longest, though, local history is short. Indian memory goes back further and would test your heart. This sacred place knew how to provide: had bear, wolves, passenger pigeons, massive trees, lakes and streams jumping with fish, lush berry patches. Traces of those times fade into the land's contour. But in a place that has been settled by Europeans for only a century and a half, farms held and worked by the same family are rare and publicly commemorated.

Somewhere between bear habitat and brokerage houses, between unbroken woods and suburbia, we should be able to conjure a reinhabitory vision here, a design for living that would suffice for hundreds of years. We should be able to constrain our wants, supply our needs, and restore much of the land to its wilder ways. Farming and forestry should rightly be the greater part of the plan. They go with the place, ground us.

The morning after Thanksgiving a farmer delivered some straw bales for me to use, first as winter insulation around the base of my writing studio, then as summer mulch. This was an energetic character, a seventyish descendant of the first settler in the township. He's an enthusiastic waltzer and polka dancer. His great-grandmother, a physician, rode the 15 miles to her office in Traverse City on horseback, not in a buggy.

Over a neighborly cup of tea and piece of pie, I learned that he was still working the farm his family had owned since the early 'teens. By selling wood and maple syrup they managed to hold on to their land during the Great Depression. I also learned that it's wasteful to tap a maple tree too high because the tap stains the wood black. Along with such practical intelligence, he also rolled out a few anecdotes demonstrating him to be a driver of hard bargains and a good-humored guy, but no one to mess with.

As I listened to him and studied his round, firm visage and piercing blue eyes, all I could think of was the difference between the life of his household (he made several glowing references to his wife's womanly accomplishments—an intricate quilt sewn for a granddaughter to take to college, her home-canned venison mincemeat) and that of mine, of his working knowledge of country ways and my vicarious appreciation of them. My tenure here is a function of recent choice and romantic principle. His is his heritage. It's unsentimental, involving a lot of hard work and iron thrift. However differently, we are tied to this place. Here and no place else was where we could meet as we did. He informed my sense of place and I'm an addition to his.

With luck I may have a few more decades to devote to the problems and pleasures of living here. If I keep my eyes open and my mind engaged, my local knowledge is bound to grow and with it, I hope, some fortitude. I expect to see more change in my lifetime than that farmer has in his, with a possible return to the kind of life he knew growing up on a farm during a time of economic disarray, dug in fiercely, cleaving to the land. What I hope is that in the future the civilizational errors bringing big change down on our heads will be seen for what they are, and that our community will get beyond this mystique and further into the work of reinhabitation. For the place we live in is real, and the time for belonging is now.

TOUGH LITTLE BEAUTIES

(2001)

When we think of rarity and value, precious stones—jewels whose glittering beauty is condensed and crystalline—may come to mind. Jewels are scarce and close to indestructible; hence their worth. Fragility also can render things rare and precious. A rare thing is one of which there are few, although that rarity may once have been abundant, even common. Human activity has effected the rarity of many once-common plants, animals, and kinds of natural communities.

French Bay State Natural Area on Beaver Island in northern Lake Michigan has abundant arrays of the rare dwarf lake iris *(Iris lacustris)*. From the iris, I learned that fragility is situational. Although the plant is regarded as a threatened species by both the State of Michigan and the federal government, these bright, beckoning inch and a half tall irises are both fragile and not.

They flower around Memorial Day. Then these pale lavender irises with white blazes leading to stubbly saffron beards look like a welcoming scatter of stars upon the conifer duff. Jade fans of scimitar-sharp leaves outnumber the flowers and denote the iris rhizomes' presence.

The blossoms won't wait, and the iris won't grow just anywhere. It was both necessary and a rare pleasure, at the end of May, to take a break from another project to fly to Beaver Island to see the irises in bloom.

During the last glaciation, *Iris lacustris* probably migrated as far south as Kentucky and Tennessee. They now live in coastal forests on the shores of northern Lake Michigan and Lake Huron in a narrow zone of dynamic extremes at the margins of limestone gravel and cobble beaches. The irises require the partial shade of the cedar, hemlock, and other evergreens growing between the open water and the hardwood forest a little further inland. To flower, they need some, but not too much, direct sunlight. Calcophiles, the irises also need the limestone chemistry of soils that derive from this northern Michigan parent rock. What's more, they need the lakes' tribulation: "Lakeshore disturbance in the form of wave action, ice scouring, fluctuating water levels, and wind damage to the adjacent forest edge," write James Van Kley and Dan Wujek in a monograph on the iris, "may help maintain the openings, small trees and thin soils essential for *Iris lacustris*."

These threatened plants are tough enough to have endured through the years up near the straits of Mackinac: tough enough to endure wind and ice and blowing sand, to survive under the snow, and to cope with the fluctuating lake levels and changing precipitation. They've occupied their niche and it has conditioned them for millennia.

In their natural habitat, dwarf lake iris seem plentiful. During their brief moment in the sun they're certainly the showiest of wildflowers. On that day at French Bay, my companions and I noted plentiful swaths of iris leaves and flowers by the path that led down to the bay. We spotted them back under the low-slung cedar boughs in that shoreside forest where the light could angle through, and in places a little farther from the beach's edge, where a gap in the canopy might let the sunshine fall.

The walk to French Bay takes you down from the level of ancient Lake Algonquin, which existed about 11,000 years BP, toward today's Lake Michigan. We threaded a narrow trail down a twenty-foot bluff, admiring the abundant jacks-in-the-pulpit, wild sarsaparilla, Solomon's seal and the numerous delicate sedges, ferns, and mosses of the forest floor. Along the way to the shore were swampy, tea-colored ponds, their dark surfaces perfectly mirroring the enormous leaves of skunk cabbage. White birches shone among the grey trunks of beech, maple, hemlock, and their equally somber-barked associates.

In this emerald realm under the hardwood canopy, Black-throated Green Warblers called "zee zee zee zoo zee" and black and white warblers, "weeza weeza weeza;" their songs mingling with those of pewees, wood thrushes and countless others. Breeding season bird surveys on Beaver Island have found almost ninety bird species.

When you arrive at French Bay, all the colors change. There's a long, bleached sweep of pebbly shore; an acid-green, algae-rich slough; a sand-and-gravel bar exposed by falling lake levels. Shorebirds skitter on the bar and tadpoles shimmy in the warm soupy water left in the slough. Visible across Lake Michigan's deep-hued waters, other islands of this archipelago float, forested.

More than once, I've heard oldtimers say that their folks thought the woods would never come to an end. Like most forested places in the Lakes states, Beaver Island was cut over a few times. Sure enough, the trees have grown back, but with each cutting, some richness of the forest has been lost.

It takes more than three or four kinds of trees, trout lilies, and trilliums to make a Great Lakes forest. It takes a rich and fitting community of wildflowers like the irises, violets, polygala, wild lily-of-the valley, and starflowers. It takes stumps and root wads and tip-ups, blowdowns and widowmakers. It takes fallen tree limbs and trunks crisscrossed like jackstraws to structure the niches for all the animals necessary to the forest's flourishing. The habitat preferences of the many bird species can be as various as their songs and plumage, ranging from treetops to cavities in snags to quiet places on the forest floor.

To maintain a water regime that can sustain a diverse Great Lakes forest, it takes mosses and ferns and unbroken canopy. It takes the kinds of ants that plant some of those wildflower seeds, takes the squirrels to plant the heavy-seeded trees and even wood turtles to plant mayapples. All unseen by aboveground eyes, it takes webs of fungus—the *mycorrhizae*—to connect the roots of the forest plants and enhance the subtly linked plant community's ability to take up nutrients from the soil.

It takes uninterrupted millennia to grow a real forest on a sand dune–covered hunk of limestone in Lake Michigan, and only some form of natural areas protection can secure the time for the forest to continue. One hundred and fifty years ago, the old growth forest communities around French Bay would have been nothing special. More of Michigan was natural than not. Soon enough, there was woodcutting at French Bay, as in most places where there were woods to be cut. Moss and lichen-covered stacks of cut cedar dating from the wooding days rest low and linear a little way back from the shore. Those firewood stacks, now resembling old stone walls, bear witness to that cutting.

Logging is drastic, but even light visitation can undermine the integrity of wild places. Spotted knapweed, a ruinous invasive species, was growing on the beach at French Bay. There are only a few knapweed plants there so far, but ecological invasions can begin with only a few alien organisms. Like other visitors, I may have carried in a few extraneous seeds on the soles of my shoes and served as another vector. The innocent human potential to damage nature is difficult to own.

There's nothing natural about a crowd of tourists, however respectful and careful each one of them may be. It's a good thing that there wasn't a well-marked trail to facilitate streams of visitors to French Bay. The dwarf lake iris aren't fragile, except that they could be quietly loved to death or, worse, annihilated by the clearing and soil disturbance that inevitably accompany lakeside construction. What is fragile, then, is the integrity of the habitat and its natural community as it has developed over time.

Never-ending work in preserving and protecting places like French Bay—the once-common places that have become rare, precious, and fragile—means that there may be a future for all those tough little beauties like the dwarf lake iris. The existence of French Bay, and of Michigan's natural areas means that we, as a people, are showing respect for other life communities, showing the restraint that is a sign of decency. We can afford to let those places be as they are and protect their integrity. And this means that the future might just include a new abundance of sweet old beauties that passed through a perilous time of being threatened and rare.

REVERENCE FOR FORESTS, REVERENCE FOR WOOD

(1999)

Where I live in northern Michigan, enough scraps of woodland remain to play out the yearly eastern forest pageant: the transformation from winter stark to tender spring to summer lush and dazzling fall. I grew up in sun-scourged Phoenix, an irrigated oasis surrounded by the cacti, ocotillo, and mesquite of the Sonoran desert, but the four seasons of Michigan, however cold or wet, and Michigan's deciduous trees suit me fine. Learning about my adopted landscape has meant learning a forest history. Studying early land surveys and soil maps, I've learned that this peninsula once was lushly forested. This region, which was doubtless gloomy and impassable, swarming with blood-sucking insects in summer and resounding with the howls of timber wolves in winter's depths, belonged to the great North Woods.

In Michigan, as everywhere in North America that was not prairie, desert, or tundra, there were trees enormous beyond our ken. Here stood white pines well over a hundred feet tall, hemlocks half a thousand years old, and sugar maples with trunks five feet across. Such trees constituted a bonanza of lumber, tanbark, shingles, and fuel for early American and European settlers. Within the few hundred years since settlement, the primal forests of the North Woods have been diminished to a tiny fraction of their original extent. Remnants of wild eastern forest of any type, whether pine, cypress swamp, oak-hickory woods, or the maple-beech-birch-hemlock forest of the Great Lakes states, are scarce and in as urgent need of preservation as any other wild forests in the world.

The American and European settlers who came to northwest lower Michigan in the mid-nineteenth century brought their axes and saws. They logged off the forest; then used what was bared for farmland. The hardwood timber fueled the steam vessels that plied Lake Michigan or was shipped south to put maple floors beneath Chicago feet. Today we have a patchwork landscape of fields, suburbs, third- or fourth-growth deciduous woodlands, and monocultural conifer stands planted like row crops to stabilize the soil or to harvest as Christmas trees. It took me a few years' residence in northwest lower Michigan to realize that this pleasant surround is, in fact, an ecosystem out of whack.

My ignorance concerning the native vegetation was not atypical. Most of us moderns grow up confused about what goes where botanically. The last few centuries' enthusiasm for moving plants across continents, over mountain ranges, and across oceans has sown confusion. This vast rearrangement of the world's flora has been and is driven by a quest for novel ornamental plants, fast-growing trees for forestry, vigorous forage for rangelands, and new food crops. If a weed is "a plant out of place," then most of the world's landscaping, horticulture, silviculture, and agriculture consists of the propagation of weeds. As the current steward of thirty-five hard-used acres, I dream of a modest measure of ecological restoration through easing out the non-native Scots pines that dominate my land and encouraging the return of its original community of deciduous trees and woodland plants.

In mundane matters and by the action of my soul, I try to show reverence for forests and reverence for wood. By protecting my small expanse of regenerating woodlands, by planting native trees and removing the alien Scots pine, I hope to recall some land to health. In my dealings with trees, animals, and soil, I am courteous and grateful. Sometimes I make a ritual offering of tobacco to my giant backyard beech tree. My [wood-framed] house is built snugly enough to husband the heat the stovewood yields. Paper being tree flesh, I am as sparing in my consumption of it as I am in my meat-eating: not abstinent; not casual.

Even without much help from me, the quaking aspen, fire cherries, sugar maples, and beeches of the original forest are resprouting wildly amidst the alien pines in their serried rows. Other trees and the wildflowers that belong here have a rougher go. The too-numerous white-tailed deer that browse the woodland understory fancy native yew and hemlock treelets but disdain Scots pine seedlings. Their grazing does not help my efforts to bring back this primal forest. Meanwhile, an ocean away, volunteers working with an organization called Trees for Life aim at restoring the Caledonian Forest in Scotland and plant Scots pine seedlings. They have to protect their plantings from browsing deer, too. Such are the ironies of having plants out of place.

A few relict clumps of hardwoods appear at the edges of my acreage. Among them is a fabulous beech, probably a "witness tree" left standing to denote a property boundary when the land was cleared. Unlike its columnar forest-grown sisters, this tree has, for its seventy or eighty years, stretched out its limbs to luxuriate in Michigan's none-too-plentiful sunshine. The resulting torque of trunk and expansive branch-dance make this beech a veritable *Fagus religiosa*, a leafy chapel wherein I may seek enlightenment and meet the occasional porcupine.

I believe that this beech, like every living creature that I encounter, has something vital to teach. Old-growth forests are Gaian superorganisms. As communities of beings from charismatic macroflora to obscure microfauna living together through time, forests exemplify the sublimest qualities of life on Earth: interconnection, dynamic equilibrium, diversity, resilience, particularity, adaptation, endurance, mortality and renewal, intricacy, and beauty.

The loggers and settlers of these parts were so thorough in their efforts that to stand within an old-growth hardwood forest today is an exceedingly rare experience. I have done so, though, and I now know firsthand that from mites and moonworts to all the flying, crawling, snuffling inhabitants of standing snags; from shaggy leaning yellow birch to iris-fringed tarns, there's a wholeness and holiness in an ancient eastern forest that should not be suffered to vanish from this Earth.

Around here a century ago, thousands of square miles of such forests became fuelwood that disappeared up the stacks of steamboats and smelters. Today's thirty- to fifty-year-old hardwood trees are logged to make shipping pallets that will be used once and then discarded. All around the world, wild and regrowing forests are clearcut to make disposable diapers, phone books, and daily papers. Given that woods and forests harbor much of Earth's remaining biological and cultural diversity, it seems a paltry fate for them to provide throwaway commodities for human consumption. It's like razing the cathedral at Chartres and crushing it for road gravel. Quite as urgent as preserving wild forests, restoring ecosystems, and practicing sustainable forestry, then, is minimizing our consumption of trees. Luckily, just in the nick of time, the means to tree frugality are emerging: tree-free paper, alternative building materials, and housing design that is both parsimonious and convivial; all of which evince a fitting respect for the forest.

A society premised on reverence for forests (and all other natural ecosystems) would generate livelihoods from the land. Nature would set the standard of practice and of living. Heeding nature means an unobtrusive harvest. Wood from mature trees is precious as jewels. It should be used only for the worthiest purposes: furniture, musical instruments, ornamental carvings. Such "reverence for wood," to use Eric Sloane's phrase, might help humanity reclaim time-honored skills and regain a natural aesthetic. The third- and fourth-growth woodlands of today could gain ground and, with sensitive tending, someday undergird their local economies.

Living like there's no tomorrow is guaranteed to make tomorrow no place you'd want to live. In contrast, reverence for forests and reverence for wood might make a future of dignity and delight for posterity and for the myriad beings with whom we share the planet. In wooded bioregions, regenerating forests would be integral to an economics of place and permanence, a way of life where human cultures make common cause with native ecosystems. With forests to teach us, we might enjoy a restored sense of continuity, of just proportion; we'd understand the interplay of light and shadow and take refuge within the mystery of Life. Through the woods, we might find our way back home.

Paper Birch

(1990)

As her emblem, the writer wanted an image of the beech. She wanted to pay homage to a great beech tree where she lived and, frankly, hoped to invite some of the beech's totem power into her work. In Donald Culross Peattie's *Natural History of Trees* she had read that "Our word *book* comes from the Anglo-Saxon *boc,* meaning a letter or character, which in turn derives from the Anglo-Saxon *beece,* for *beech.*" Scribing signs on the blue-gray smoothness of beeches is a very ancient pastime.

The writer approached a friendly artist, thinking that perhaps she could trade some of her work for his and get from him a drawing of a beech leaf, nut, and bud. As a matter of fact, the artist said, he was making a book that would combine his drawings with short pieces by various writers. Would she like to write to one of his images? Yes she would. This seemed to be a perfect trade and fun, besides. She went to the artist's studio to look at his pictures.

There were magical night scenes and surreal landscapes, all finely drawn and delicately colored. Considering them, the writer was at a bit of a loss. Intuiting this, the artist drew her attention to a tinted etching of a section of a paper birch trunk, which perfectly evoked that tree's calligraphic bark with its subtle undertones. In another work, a painting of many paper birches together, curly and peeling, the artist had secreted a piece of bark. He challenged the writer to find it. His artifice was subtle: so much so that after a while she abandoned her attempt and departed, letting him return to his work and returning to her own.

The painting and etching had put her in mind of the paper birches growing by the roadside that she had noticed during a recent storm. They were clad with wet snow, the delicate garnet of their uppermost branches incised with white against the bosomy gray sky. Driving by that late February afternoon, she had thought she detected a faint pulse of spring in the rosy semblance of warm flesh under the silky, pale surface of the birch bark.

Birch trees have been recorded as growing to be eighteen feet in circumference. Assuming they could find a foothold, it would take four or five Druid scientists to dance around such an elder. The writer imagined the time, not so long ago, when her land was deeply forested with birch and hemlock, maple and beech, great tall trees and their many companions: blossoming, buzzing, and caroling. She imagined a place where ancient trees were not vanishingly rare.

A story tree like a champion birch or a great hulking beech is not just any old elder, but a being who manages to live marvelously long, becoming more wise and benign decade after decade. A grandfather birch grows more lined and papery and pale, exploring, through the passage of the years, his changing essence until one day his time for living is done. He dies, quietly, turning soft and dry within, and stands for a while as a refuge for birds and insects. Then one day a gust carries him down to the ground. With beetles chewing and fungus threads questing, the grandfather subsides back into the earth. The chattering of the wind in his leaves is long gone, bright truths all carried off on a dying breeze.

––––––

The morning after her visit to the artist's studio, the writer went walking on her land with a friend who was to share his knowledge of tracking. Their heavy-booted progress through an overgrown plantation of Scots pines was slow and easy, reined in by the deep snow and their purpose of seeing. Where footprints showed that an animal had passed close by the low branch of a maple sapling, the tracker showed the writer how to read the creature's other traces. Studying the little tree, she discerned a fine hair clinging to a twig. Her breath caught the wisp and lofted it away.

The friends walked on and emerged from the pines and then ascended a slight rise to the grandmother beech. To make a venue for a practice of sitting and watching, they trod down a spot in the snow, spread out a neat waterproof groundcloth and sat, separate and still. Akimbo shadows of beech shoots became sharp, then vague, on the pristine surface of the snow, according to the waxing and waning of the sunlight with the passage of filmy clouds. The writer listened to the rattling of the branches of a nearby sugarbush in the freshening breeze and was sure she heard there another portent of spring. At length the two walkers rose to return, this time zigzagging a different route through the parallel rows of exotic conifers.

Nature has always gone to marvelous lengths to thwart desolation and sameness. The irrepressible hardwood forest so testifies. Popple and cherry, young maple and beech—its outriders—assert themselves at every opportunity, and its other, more reticent, members follow on. They thank the pines for their service in securing what remained of the soil after the land was cleared and tilled and hasten to enliven the manmade monotony.

Thus it was not all that improbable, but a graceful coincidence nevertheless, for the walkers to chance upon a youthful paper birch gleaming in a clear place. The writer couldn't recall ever having seen it before. The discovery was wonderful to tell. Alone that afternoon, while the west's glory dwindled, the writer strove to put characters together to detail the apparition of paper birch. She mused long into the dusk, envying the slow, calm artistry of trees.

IN HER WOODS

INDIA JOURNALS

India Journals

In September 1992, I made my first visit to India. An invitation
to participate in a conference sponsored by the Ladakh
Project—"Rethinking Progress"—provided the occasion.
Since a journey to the East is no trivial thing, I decided to stay a
few weeks and see a little bit of the complex, disparate country
that is India. I flew via Tokyo to New Delhi, where I stayed at
a YMCA Hostel. One day later, a quick flight transported me
and my fellow-conferee Gary Snyder and his son, Gen, over
the Himalayas to Leh, capital of Ladakh (altitude 11,500´).
The conference and related festivities transpired over a week.
I returned by air to Delhi to embark from there, by train, for
Varanasi (Benares). A delayed departure allowed me a day's
tourism in Delhi.

After two days in Varanasi, I traveled south two days by
train to Madras. After spending a day there tending to some
business and being a tourist, I boarded a bus south again for
Pondicherry, a small, affluent, coastal city, which had been
the capital of a French colony. Pondicherry also was the home
of Sri Aurobindo, a 20th-century philosopher-sage and hence
is a center of activity for many of his devotees.

I went to Pondicherry in order to visit Auroville, an
international intentional community nearby. Auroville
was established in 1968 at the direction of The Mother, a

remarkable woman who was partner to Aurobindo for many years. Her vision was to establish a township for all of humanity, an "experiment in human unity," and this task included ecological awareness—reforesting the denuded lands where Auroville was founded—all 2000-plus hectares.

This "Greenwork," heroic in its scope, was what drew me to this extraordinary settlement. As luck had it, I arrived at the beginning of a conference, "Awareness Workshop for a Sustainable Future," that was being presented by Auroville's Center for Scientific Research. Topics included afforestation, water resources, wastewater recycling, sustainable agriculture, earth architecture, ferrocement, and windmills. After a few days in Auroville, I took the bus back to Madras, rested for a day, and flew to Delhi where, for a night and a day, I enjoyed the hospitality of a new friend, Rukmini Sekhar, who was another participant in "Rethinking Progress" in Ladakh. From Delhi, I returned by air, via Tokyo, San Francisco, and Detroit, to the Cherry Capital Airport and a quiet, comfortable life whose inner dimension has been subtly churned by fleeting glimpses of the sights and sounds, the faces and metaphysics of places in India.

What follows is just journal-writing. It's partly a diary but, more than that, it's conversations I had with myself—keeping myself company, keeping me sane—as I made my way along a journey of a thousand miles and thousands of years.

TOKYO. SEPTEMBER 2

I'm at a hotel near Narita Airport, which serves Tokyo. The sun is rising over a dark stand of young pines, over small plots of taro and onions and corn, over smaller versions of supermarkets and malls, over another country. I'm in Asia.

This overnight stay in a squeaky-clean airport hotel afforded a chance to wash and sleep and eat in a generic form of luxury. The flight crews stay here. It seems to be no place in particular. Narita Airport is an armed camp, a militarized zone. There's a double row of outward-curving chain link fence and concertina wire. There are riot police on duty at all times. Security is very heavy because of the ongoing peasant resistance to airport construction.

Things are pretty and sensible in this hotel, if sterile. If everyone on Earth insisted on this, we'd all be goners. The room is air-conditioned to a chill. There are all sorts of inviting throwaway packages of this and that—soap, toothpaste, shampoo. Lazily, I availed myself of the goodies. It was easy to do. I took a wonderful hot bath. The water was pale azure in the tub; somehow it looked different here. Before the departure of the bus that would take us back to Narita to board the flight to Delhi, I was drawn out of the hotel by the sight of a tree that looked like a beech, trunk- and branch-wise, but more like an elm in the arrangement of its leaves.

It seemed important to set foot on Japanese soil, so I went for a walk, saw startling things—galvanized shacks by the dirt footpath, falling into disrepair, a naked pink baby doll upended in the trash. Another structure of galvanized tin, but this one a home, complete with a postbox on the fence and sounds of family emanating from within. I walked among young stands of cedar, an abandoned chestnut orchard, saw huge insolent cicadas and great lanky yellow and black spiders ensconced in yard-wide webs.

Is it still Wednesday? Whatever day it is, I'm in New Delhi. It's so hot that it's like not having skin. A place the temperature and humidity of blood. I believe I'm one of the ones who loves India at first sight. A great big Ganesha, the elephant god, flanked by shoulder-high brass candlesticks, greeted us passengers at a switchback in the walkway from planes to customs.

Everything's in some degree of disrepair, nothing works altogether dependably. Lack of funds dictates the relaxation of certain standards of slickness. People aimlessly sweeping, mopping, brushing around small accumulations of trash. There are rubbish heaps, but not of the magnitude that we know in our cities. Lovely women, princely men, limbless beggars. Pandemonium of horns and motors, traffic grinding loud, frantic, relentless. "Madam" and "Hallo" dogging my passage, no opportunity to linger and observe invisibly, no chance of not playing my role as a western tourist. A rupee, worth less than a nickel, is a significant sum. The beggars touch their foreheads when and if they have received alms. There are a million little enterprises. Vegetable vendors, ragpickers, buyers of old newspapers all come to urban enclaves making their distinctive cries. The only place in the U.S. where you would hear such a wild diversity of hawkers' calls anymore would be at the circus.

Checking out of the YMCA hostel this evening brought several procedural strands together in what seemed like a tighter and tighter knot which at the penultimate moment of aggravation, suddenly untied itself. "You decide," said the clerk, who had just presented me with an unforeseen option, smiling implacably, gently bobbling his head, knowing completely. In the midst of frustration with the foolishness and impenetrability of all the form-filling, an access of love. These beautiful, civil, intelligent, enigmatic people with their carbon paper! Then an odd little shift, from my petty antagonism over to good will, the dawning of a shared sense that this is a funny game we're playing and why not play it? All will eventuate as it should.

Ladakh. September 4

It's 3 a.m. Friday in Little Tibet. Outside there's the sound of rushing water, breezes perhaps, and a cacophony of barking dogs. Electric lights beaming out from the monastery up on the ridge and from this guest house enclave, streaking the fields and stone walls with rude illumination.

After breakfast, Rukmini Sekhar and I went for a walk up to the *gompa* (monastery) along a concrete footpath edged with rushing streams and walls of stones accumulated over the years, past fields of ripening barley. These footpaths are the thoroughfares and the community is in motion. A cluster of women stopping for tea in a corner of a field, a monk returning from a garden with kohlrabi in one hand and asters in the other—food for the body, food for the soul—a little boy bearing a heavy load, a bale of alfalfa trailing a strand.

Lots of rock-raw slopes rising up all around, ridges crowned with monasteries and *chortens* (Buddhist monuments). Huge prayer wheels, big polychrome cylinders. You spin them and they ring a bell. Done by you, for all sentient beings, prayers cast to the winds.

LADAKH. SEPTEMBER 6

It's been overcast the two days we've been here. John Paige and Helena Norberg-Hodge (our hosts at the conference) say the weather is changing. An ominous possibility. Plenty of motor noise down in Central Leh, now a loudspeaker. This morning, before dawn, a melodious call to prayer from the mosque. On the path daily, a man hauling bricks on his back, using a tumpline to carry the load. Ladakhi people are handsome, their picturesqueness varying as to the degree to which they have adopted western dress.

The inventors and promoters of disposable packaging should find a special place in hell. There's a stretch of the path down to town where the stream opens out and braids through a little greensward. People go there to wash and bathe. Also to shit, apparently. There's litter all over the place, hung up on cobbles in the stream.

For Helena, watching this degradation befall Ladakh must be like watching one you love suffer a life-threatening disease; to see the ravages, subtle and gross, of progress, and to be negotiating compromises, attempting to hold the line against total destruction. On my arrival I had been, in polite guest-fashion, exclaiming over the beauty of the place. Helena said she thinks Leh is ugly. Truth-speaking. Electrical wires and smoke-belching jitneys and disheveled western dress are as unbecoming here as anywhere.

This is the first time I've ever experienced or, rather, *observed* a village. Very lively, very like an organism, very much a world of interpenetration, all ages and occupations in a swirl together.

Very much a sense of a common life. Even in the small territory that I have moved within these past few days, it's seemed like people are not private about their lives, or rather that much of life, in the summertime at least, is lived in the open.

Across the way a man is heaving bunches of alfalfa from one level of his roof to another, where it will be set out to dry. There's a quantity of smoke billowing up from downtown. The yap dogs (Lhasa apsos, native here) are yapping; full din of horns in bustling downtown Leh, into which I ventured this afternoon. There are a number of restaurants and hotels and souvenir stands. Lots of tumble-down buildings, a high quotient of funky, scrawny, resourceful dogs (Helena tells us that these dogs arrived with the beginning of waste. There was no garbage to support them previously).

In the small plot opposite, a group of men and women are harvesting grain. They're singing, a bell-like song with an antiphonal response. Crouching, using sickles, they move across the field felling the pale gold stalks, disclosing the dark yielding earth. A song that continues, the singer sparing breath from his labor to add music and grace. It's hypnotic, a sustained rhythm and melody that weaves their work into all of life, rather than consigning it to the category of toil. Oh, may this culture persist!

A big event last night. Women from all over Ladakh came to a party dressed in their finery, which is fabulous indeed. A few of the women were wearing their turquoise-and-coral paved headpieces, and most of them were wearing beautiful necklaces of seed pearls and gold. We were served butter tea, a beverage made of water, green tea, soda and yak butter. I consumed a fair amount. It wasn't too bad, and if I hadn't been feeling so protective of my digestive tract (and with very good reason), I would have consumed it more freely. It's very rich. We were given apples and grapes, seed cakes and chapatis, and a tasty stew of potatoes, peas, and perhaps goat. Then a bowl of apricots in syrup for dessert. We all left about nine and learned that the party, and the meal, had hardly begun. Short hitters from the lower elevations.

Each day here is like a small lifetime. Yesterday we were taken sightseeing. The philosopher Tashi and the lama Paldan accompanied us. We drove in a caravan of taxis over to the Indus valley. First we saw some shallow bas-reliefs of various incarnations of the Buddha carved on great boulders perhaps 30 feet high. Very old, a millennium, perhaps.

Then we visited a monastery which houses the largest (40 ft.) statue of the Buddha in the whole region. The gompa was perched high up, as they all are, and was partly vital, partly disintegrating. There is a tumble-down quality to a lot of Ladakh. Many of the structures are old, built with available materials—by vernacular architects: frugal, sensible, ingenious. Bespeaking tremendous labor.

Women were hauling baskets of sand up the long path to the monastery. Men were doing the carpentry, monks supervising. The iconography on the walls of the courtyard, shrine, and meditation hall was vast and lively, strange colorful images to elucidate Buddhist teaching.

Paldan and Tashi provided commentary and explication of the images. Paintings of the life of the future Buddha covered the wall behind the mezzanine from which we viewed the Great Buddha's head and upper torso. Butter and *tsampa* (toasted barley flour) offerings placed before him, bowls of water, two English pounds.

A spanner in the works. My train is delayed until 11 AM tomorrow. I'm not at all clear how things are going to work out from this point.

India is infinite bullshit, infinite pathos, infinite beauty. Every little task is confounding, snarled in a million petty procedures and misdirections. Infinite bullshit demands infinite patience, and that I lack. Mine has failed more than once already. This is definitely one day at a time country.

I arrived at the Y after the morning flight from Leh, reorganized and checked my bags, and sallied forth to do errands. The ricksha wallah, first of many today, dropped me at an optician's, where they generously replaced the lens in my sunglasses, gratis. I headed off for the National Museum with another wallah, a fairly sleazy, toothless guy, who, as they all do, offered me a complete Delhi tour, and was very undiscourageable. Halfway down the road, I thought perhaps to buy my air ticket from Madras to Delhi and changed our destination. We wheeled around, kamikaze-wise—the traffic patterns are ineffable, amazing, hair-raising, non-linear, and surprisingly non-lethal given the anarchy of it all. This wallah took me to the airlines office, then asked me the time, then informed me that it would not re-open until two, an hour hence. "Why didn't you tell me that before we set out?" I asked him. "I have no watch," was his answer.

At the museum, there was virtually no art that could be called secular, save a few decorative works, for instance an early 20th century ivory screen produced here in Delhi. The screen consisted of dozens of carved panels about six inches on a side, pierced in a variety of geometric patterns, the openings between the ribs perhaps an eighth of an inch wide.

It bespoke a staggering level of artisanship. All that intricacy, delicacy, perfection.

But the majority of works were devotional, save the prehistoric artifacts from Harappa and Mohenjo-Daro. Tiny little pottery figures from nearly fifty centuries in the past included crude beginnings of representation—eyes made of tiny balls of clay, incised for pupils, pinched on to the head of the horned or beaked being—and marvelously faithful, original renditions of a sheep's head, a dancing girl at rest.

There was a room of bronze divinities—Shiva, Parvati, Vishnu, a half-dozen others, voluptuous graceful variations, portrayals of gods and goddesses in states of calm, infinite bliss.

There was a gallery of tribal art—at first I thought I had stumbled into some extraneous African collection of dark, gaunt wood effigies. But no, this too was Indian expression, and, in the midst of all the grotesque (but numinous) statues, a female torso of an elegance and lilt and that caused me to weep for joy at what humans can create.

The few hours spent in the museum disclosed yet another vast realm of cultural riches, dozens and dozens of genres, regional styles spanning the subcontinent: India's five thousand years at headwaters, flood tide, and broad delta of a high civilization.

Working people are very lean, skinny, and watchful—whatever's going on here, some of it the weather, probably some of it the conditions of health, is whittling people very keen. Yet they move at an unhurried pace, and there always seem to be two or three guys hanging out and chatting idly for every one man doing something.

The bank where I exchanged some currency today was purely Kafkaesque, thronged with people presenting forms, filling out forms, examining and stamping forms. No fancy interiors, no recognizable logic in the layout, a vast variety of banking activities taking place. Somehow it all works. Everything is dingy and beat-up, comfortably funky, slightly mildewed, borderline squalid. It's a real world, all manner of reality pouring in: noise, fragrance and stench, horror and grace. Beggars are real, and it took me almost no time at all to harden my eyes against them. I passed out coins, peanuts, cashews, but by the time I reached the train station, which was, as promised, intense, my heart was harder. There is such an ocean of need here, with these millions, that there's no place to begin or end.

Around the station were cows, saddhus (ascetic wandering holy men), beggars of every age and condition, terribly persistent kids, relentless and staring. Scrawny pitiful basilisks. Unsurprisingly, it was at that point in the day's travels that I had my first overpopulation thought since arriving in India. Family planning would forestall this gathering of apparent misery, or so the theory goes.

India is a great place to work on your boundary issues. I must have said a firmer no than ever before in my life at least a thousand times today. I struggled manfully to the front of the line at the railway inquiry window after an abortive first attempt, and thanks, finally, to the courtesy of one of the dozen men pressing forward to ask a question succeeded in getting an answer.

New Delhi. September 13

An hour on the platform at the New Delhi train station equals a year of life in the good ol' U.S.A. Whatever it is that is being asked of me here, it may not be charity. Charity meets with a peremptory grasping. The beggar children are relentless. It's the minimum humanity, and they are tough. I got to the station, secured a porter who was accompanied by a tout who ran me all over the neighborhood in search of information that was readily available at the inquiry window. The tout also tried to make off with my luggage but fortunately I was alert enough to see through his simple ploy. The porter was handsome. A basic tool of his trade was a cotton scarf that he coiled on top of his head to cushion the weight of suitcases, and later used to bind up his knees as he sat on the platform chatting with his buddies while we awaited the train.

So an irony of India may be that I'm coming away with sympathy for the rich, because here I am one, and everyone wants something. I'd rather that not be the big insight from this journey. Fact is, I arrived here with spending money in the amount of a clerk's annual salary. It does no good to protest that in my country I'm at the modest end of the scale.

Seen from the Train

Fields of cane, straw huts, no trees allowed to grow to any age. Pearly gray sky, red turned earth, squalor and splendor in the cities. Lush vegetation, water buffaloes with egrets perched on their broad shoulders, brick or mud hovels on the outskirts of town, women in saris carrying large saucer-shaped loads of construction rubble.

So now, at the end of a day in Varanasi. I have just returned from attending the Mahant's singing lesson (the Mahant is Dr. V. B. Mishra, high priest of the Hanuman temple, professor of hydrology & civil engineering at Benares Hindu University, and head of the Sankat Mochan Foundation, which among other things, is working to clean up the Ganges River).

We went to the Hanuman temple which is an open place with marble floors and arcades. There are shrines containing statues of Hanuman and Sita Ram, facing each other. Hanuman is the monkey god who assists Rama, the hero/god of India's epic myth the Ramayana. Sita is Rama's long-suffering wife. The temple pillars are vermilion up to an arm's height. This from worshippers wiping off the daub of color remaining after they have anointed themselves. As we arrived, devotees came forward to touch the Mahant's feet. After he had attended to his priestly duties, we ascended a flight of stairs to the roof, where a sheet of white cotton had been spread for us to sit on. A circulating fan stood nearby. The upper branches of great old *peepal* trees rose above and around the roof. A little harmonium awaited the arrival of the singing master, Chanu Lal, a leading exponent of the vocal arts of this region. When Chanu Lal-ji appeared wearing a lustrous silk kurta (a tunic) and an even more lustrous smile, the lesson began. Before the start, one of many of Mahant-ji's attendants brought baskets of *malas*—flower garlands, these made of jasmine, black-eyed Susans, roses, and an unfamiliar purple blossom. The Mahant then offered them to us. I wore mine so that its fragrance enhanced the night, which was pleasantly balmy. The moon was just beginning to wane. It was a fantastically poetic, medieval scene.

Today at dawn, Manon (a French Canadian scholar who presides over the Sankat Mochan guest house) woke me and we threaded our way through narrow lanes, brick-paved, dung-redolent, down to Asi *ghat* (a ghat is a flight of steps down to a river) to hire a boat. We waited a while among the bathers till a boatman came along with a guide. We put out onto the broad silty Ganges. We saw a kingfisher perched on a lamp post which appeared to be standing out in the middle of the river. The bird had a slightly different silhouette from the North American variety and a glint of green in its iridescent, even regal, plumage. Blind dolphins surfaced and dove, rafts of hyacinths swept by, and other vessels, bearing other tourists. Ghoulishly, I looked for the remnants of poorly-cremated corpses, but saw none.

The noise pouring in from all the households fronting on this lane, and from the activity on the street, is considerable, but not as debilitating an assault as the more vehicular noise of New Delhi or New York. Varanasi is fortunate in its narrow streets, byways too narrow to admit automobiles. So I think this may be the New Orleans of India, the Big Easy. It's riparian life, moving at the pace of the river, which is in flood now. The ghats are mostly submerged and the Ganges is, of course, the color of *chai* (tea), as the Mississippi is the color of café au lait. The buildings facing the river are made of brick and stucco, and painted in many colors—ochre and lavender, turquoise and henna. Bells and cymbals rang out from temples we passed as we were rowed along by the boatmen. At first it was just one big man, dark as an African. Then two lean younger men, one of them with elaborate paintings on his brow. Hard work. There is endless hard work being done, some of it quite unsavory. Out in the streets the cow flops are gathered up and patted on to walls to dry, to be used as fuel in cooking fires. The impression of human fingers leaves them looking ornamental—like rosettes or bosses.

The trip to Sarnath with Manon was a journey in itself. Suraj (Mahant-ji's driver), the mellow and implacable, piloting us through traffic jams in the heart of Varanasi—the usual throngs of gaily decorated bicycle rickshas, motor rickshas, etc. Roundabouts choked with traffic of every conceivable ilk. At length we emerged from the city into the district around Sarnath, which is more open, and tree-lined, shady, and peaceful. There are many Buddhist establishments there. It is, understandably, a center of study and pilgrimage. Then a visit to a chapel, which was decorated with Erté-style paintings of the Buddha's life. There was a great statue of the Buddha and an altar containing relics of the Buddha. A chip of bone, Manon said, a little disappointed by it. I can't see how relics could be anything but disappointing—charnel remnants of the hallowed. Proof that the great man existed, I guess. Makes it more valid. Still, standing there before the image of the Buddha under its grimy baldachin, I was stirred by a presence. Absolute compassion. Which, I guess, is about the same thing as infinite love.

There seems to be nothing like private life, at least for the poor. Sleeping on a *charpoy* (a wood and twine cot) alongside the street as the traffic of autos, motor-rickshas, bicycle rickshas, scooters, bikes, and herds of animals and people flows loudly by cannot be restful, yet that is the lot of many, as is bathing in puddles by the side of the road, and hungrily watching others eat.

Lots of biodegradability and recycling in this country. Sweets served on plates made of leaves, chai in little clay cups. Cows, goats, dogs, and swine all wander the streets munching trash. Labor-intensity, too. Resourceful, devout, incredibly hard-working everyday Indians stare and stare at the Western passerby. Me from another planet. People here mow lawns with hand-held sickles.

Kerb-side shops everywhere, alcoves above street level, wherein the proprietors are squatting—it's a chairless society—and getting about their business. Tiny little enterprises selling a few bits of coal, baskets of vegetables, chai, doing shoe repair. Indians look to be hugely ingenious and industrious people. There's little heavy equipment. Five men hoist a log out onto their heads to move it inside a small sawmill. By the side of the road one sees furniture workshops, mechanics' stands, whole families engaged in manufacturing in these booth-like storefronts. Saddhus (a saddhu is a holy man, usually ascetic) of every description, not all of them gaunt. In the *ghali* (walkway) in Godolia (a shopping district) there were half a dozen temples, from shoebox to telephone booth size, the divinity within one of them being anointed by a priest as we watched. The priest scowled, they all scowl, Indian women just stare. Western women, apparently, are roundly despised. All manner of staring. It is impossible not to be reproached by it.

New Delhi. September 17

On the train South. In the seats opposite, a Sikh husband and wife, whose sons last night, with their cloth-clad topknots and otherwise western dress, bid their mother good-bye by lightly touching her feet. I am the only foreigner-lady for miles around, and thus exceedingly conspicuous. "Where are your companions?" asked the Sikh gentleman moments after the train began rolling. I answered that my companions were in the U.S., where I live, and that traveling alone was my situation. So he held forth for a little while about the remoteness of the possibility that some mischief-makers might attempt to harm me. The only real problem, he said, would be the lack of a sense of security as I traveled. Delivered of this intelligence, he returned his attention to his wife, and for the balance of the trip was a contented half of a companionable middle-aged couple, sitting cross-legged on the berth and playing cards, or picnicking or reading. Sometimes she would sing softly, high little melodies.

For a while we were rolling past a ridge of what looked to be red sandstone hills. Now it's flat again, fields and trees, ponds starred with white waterlilies. Bigger trees now, a hammering sun directly overhead, clusters of workers pausing in any pool of shade. Elsewhere, isolated tenders of livestock sit under big black parasols. Back through a patch of scalped land, worn paths, sparse trees. India is a long poem. She seems to comprehend all of human history, from the Neolithic to the present, and in simultaneity. The train rolls by villagers waiting at crossings in two-wheeled wooden carts pulled by bullocks as geometric power towers strung with electrical transmission wires bestride the landscape.

No telling what horrors go on in the shacks of the least of the least. Passing by a settlement on a train I saw, or thought I saw, a man pulling down a little girl's skirt. At first I thought it might be fatherly care, but there were other males around, and that kind of intimacy with a girl children seems uncharacteristic of Indian fathers. So what was it, then?

This seems to be a pretty devotional hotel—perhaps they all are. Last night a priest was censing the images of the gods hanging over the kitchen-to-dining room pass-through with great billowing clouds of what incense I don't know but the aroma was pleasant. The waiter's forehead is striped white, others are wearing red dots in the middle of their foreheads. In the dining room this morning I was the only non-Indian, the lone woman. So breakfast was accompanied by the usual solemn stares. It might be fair to say that India abounds in stereotypes and archetypes. Gods and goddesses crowd in on one here; and rigid social roles, codified in the caste system and gender identity. Women are brides, mothers, or wicked stepmothers. Also heads of government, somehow.

A different kind of draft animal in the South. Cows with horns like sickles, their hides buff-colored. Crescent horns painted alternately red and green. Or bedecked with a ring and a bell. Or, nearing Pondicherry, both horns blue. This adornment, does not, however, signal any sentimental sparing of the whip when the time comes for the cart to make its move into the hectic traffic of the Madras boulevard.

I am a very tentative, retiring tourist. I walked down Monteith Road towards the government museum. I stopped at a mall to get another *salwar kameez* (a popular Punjabi-style women's costume, consisting of a tunic worn over drawstring trousers with gathered ankles) and at another for a *dupatta* (a long scarf, customary accent to salwar kameez). After changing into my finery, I went on to the museum, a big fanciful red sandstone building, legacy of the British. First a brief turn through the art gallery, which had some lovely bronzes, a couple of grand, sublime portrayals of Shiva Nataraja. Shiva is a major deity, god of creation and destruction, habitué of the cremation-ground. His dance brings the world into existence. And hanging up above the display cases, great portraits of British overlords. Served them right to be stifling in their ponderous uniforms.

Well, Indians had a lot of centuries to carve a lot of rock. Here and everywhere they transformed stones into gods, many of which now are static, housed in museums. Their counterparts at large are sprinkled with blossoms, ghee (clarified butter), vermilion; aroused by the clashing of cymbals, beguiled by incense and put into nightgowns to sleep.

One corridor held an exhibit of memorial stones—hero markers for him, commemorative steles depicting archers—and *sati* pillars showing the bangled arm of the virtuous woman who threw herself on the pyre rather than endure widowhood. What a dreadful thing to suffer an agonizing death on account of the lone fact of being no longer a wife.

The natural history part of the museum, a mausoleum of Victorian ilk, was pitiful. Dead, dusty animals. Taxidermy any place is bad enough, but in this climate every mammal specimen begins to look like a rat. Road kill and worse. Their example of the North American raccoon was like nothing I'd ever seen dead or alive.

The martyred corpses nonetheless gave me some small sense of India's fauna, deer in a wide variety of sizes ranging from small to tiny. Squirrels going from medium to huge. A flying squirrel about the size of a house cat. Also flying lizards and flying frogs.

So the bus ride from Madras was fine, really, but a lot. After getting to the bus yard 40 minutes early, which was a good thing, we jounced off, horning (Indian parlance for honking) all the way. During the wait, ever-present little boys offered this and that. I had one buy me some oranges and got into an idiot squabble with him over the price. And I also bought a flower garland to wear in my hair, insisting on receiving it before paying the little boy who trotted alongside the bus to deliver it from the tiny stand kept by his mother and sister.

What else about the bus trip? Fast and slow, depending on circumstances. Unglazed, barred windows admitting a buffeting wind. No AC necessary, ventilation, laden with diesel soot, much appreciated because of the heat. I was on the sunny side. A mother, child-woman and grandson sat beside me for half of the four-hour journey, me feeling a little guilty because my luggage took up so much space underfoot. That being a blatant metaphor for Americans on the planet.

So many grace notes in this place. Blossoms and shit. Women of every estate (or nearly every one—I reckon I haven't seen the worst yet) brilliant in their saris, earrings, nose jewels, flowers. Enroute to Pondicherry passing by a green, acid-green field where work was going on, rice being tended perhaps, a regularly spaced line of dark sari'd women bending to their task, looking like a strand of jewels.

Some huge logs on lorries sharing the road with us. Maybe not so prevalent as logging trucks in Leelanau County, but in a land hard pressed by its burgeoning peoples, trees of this size—perhaps a yard and a half in diameter—must be exceedingly rare.

The ricksha wallah was ready to pedal all the way to Auroville (a distance of about ten kilometers) before I stopped him. More often than not, these dealings are a hassle, and this was no exception. Brings out my inner termagant, a shrillness activated, I think, by fear. But it is ugly and I wish it weren't a part of me.

AUROVILLE. SEPTEMBER 22

I've been domiciled in an elegant, well-made guest house, maybe 15´ square, with windows shaped like inverted triangles, striped with green steel bars, prevention against theft. One keeps one's things locked up in India.

It's cool just now—a sweet, refreshing moment after some truly astounding heat yesterday. This morning I awoke to a glorious variety of bird song. The sound of healing in Auroville. This is the country, and it's so much more amenable—and richer in a diversity of life forms—than the city. There's a huge banyan tree at the center of the complex. What a survival strategy! A bird shits the banyan seed on top of a palmyra and eventually the banyan envelops the host tree with its roots, killing it. The banyan branches out laterally, sending down roots from these branches which become daughter trees. They can create whole groves whose progenitor is obscured by time.

Hard to put my finger on the mood here. My own mood is too dense a filter. Solemnity in the man Joss who spoke about the Greenwork and gave the slideshow yesterday. A faint tinge of somberness in Ed. Tineke seems alternately good-humored and overworked. Likable, withal. She was good enough to show me the civet kitten she was nursing. Little scaper had tumbled out of a palmyra tree. It resembled a cross between a fox and a possum. (No ratty tail, though).

Whatever is going on here psychologically, the work is good and they are accomplishing beautiful things. Aurovillians were planting their first trees at the same time I was graduating from Mills College. Much good can happen in 25 years.

En Route Home

People in India move with a sense of endless belonging, as if their presence in a particular place could never be subject to question. They look out at you. Their life on the streets—the chai-pan-fruit-snacks-barbers- and cobbler stands; the nappers by the roadside, the families camped with lunchpails and thermos on the airport kerb, awaiting—all of them assert bodily that public space is their place.

The night before my departure amid the pan-chewing glee (pan is a bundle of spices chewed after a meal) at Ruki's dinner party, my Indian friends were exhibiting a love of being Indian and doing and sharing things Indian—they had delight of a sort that I never feel over being North American. The Madrasi bookstore proprietress and her odd Aussie customer last week expressed the world's hope for George Bush's defeat. Sobering thought, that—that on the far side of the earth people flinch at the possible outcome of a U.S. election. Too much power, too little goodness. What a terrible path my country has been led down. Distracted by the Cold War shibboleth of Communism we became the greatest secular materialists of all. I'm not real happy to be an American, and certainly not proud. I really wished for a recognizable name to call my homeland other than the U.S.—the Great Lakes Bioregion doesn't yet ring any bells. We are so brash and terrible a nation-state, and yet we are a people of awesome privilege and possibility.

So good-bye to India, land of many odors and fragrances, terminally funky, fantastically inefficient. India, which enjoys the protection of three hundred million gods. The ricksha wallahs have foil stickers of Shiva on their dashboards (also, on their mudflaps, bold paintings of a person who appears to be a hybrid of Elvis and Rambo). India, which bodies forth saints. Gandhi's portrait is seen everywhere. India, where when a waning old man on a train mutters "Ram Ram Ram" on stirring from his sleep, he's not calling for booze, but keeping the name of God on his lips. This is a great country, a great people. I bless the fortune that allowed me to visit this overwhelming land.

THE BODY POLITIC

On Fetal Personhood

(1986)

(Written in response to a question posed to
CoEvolution Quarterly's *regular contributors)*

In answer to the question of what would happen if we treated
the fetus as a person, it seems obvious to me that we already
do: those we favor, we nurture; and those we don't, we discard.
Human beings always demonstrate a meanly uneven regard
for the worth of a person, and that unevenness depends upon
our varying attitudes about sex, class, race, and timing.

The question brings to mind Florynce Kennedy's sharp
comment, "If men could get pregnant, abortion would be a
sacrament." If men could get pregnant, they wouldn't be men;
if there weren't two sexes, with all their attendant differences,
who knows? Maybe the tendency would be to treat all persons
equally. But as it is, motherhood is a determinant of women's
experience, for better and for worse.

Maternity is doubtless a primal spring of compassion and
tenderness for the human race—small wonder that "God" was
actually understood to be "Goddess" until relatively recently
in the history of the human species. In our time, however,
maternity is the reason that the vast majority of the poor are
women and children. Reality is that this society, like most
others, is not much inclined to treat fetuses, children, or
women as persons.

Judging from the incidence of default in child-support payments, a considerable number of fathers are not prepared to treat even their own offspring as persons. So if the educational task of getting all women to regard all fetuses as persons would be formidable, I dare say the challenge of getting all men to treat all fetuses as persons would be just about impossible.

Would it be possible to treat all women and all fetuses as persons simultaneously? Not unless it becomes possible to convince all women that carrying all pregnancies to term is the right thing to do, regardless of circumstances. So what if some women abort for selfish reasons: most abortions are altruistic and most women know that we live in a world where people aren't even treated as persons; so to give birth to a child who may want for love or money or other support is not a kindness. Therefore, women who are presumptuous enough to believe that, as mothers, or would-be mothers, they are in the aptest position to make quality of life judgments for their offspring will have to be reeducated. And what will they be told? That on a finite planet, wishing can make it so?

The process of convincing women to believe that their personhood should be subjugated to fetal personhood whenever that is demanded could be a fierce challenge. There are virtually no societies without abortion. Women in every culture are willing to go to great lengths to attempt to space their childbearing and limit their family size through whatever means available, notwithstanding patriarchal metaphysics that proclaim abortion to be wrong.

I've never known anyone to have an abortion casually. Most of the pregnant women I've known, women who sought and welcomed pregnancy, engaged in ongoing conversations with their fetuses, whom they choose to regard as persons-to-be dwelling within them. It appeared to be a touching and potently educative relationship. The women I know who have had abortions (the same women, in some cases) did not really treat their unwanted fetuses as non-persons, either. Grief and loss, foregone possibility, were accepted as the mortal consequence of the choice to abort.

Abortion, even as a widespread practice, is a very intimate act. It is a taking of life, but a taking that is personal and knowing. Abortion has gone on alongside the giving of life for as long as women have been able to accomplish it, safely or unsafely.

It is not for nothing that the aforementioned Goddess was understood in her triple aspect: virgin, mother, and crone, old "harry me, marry me, bury me" herself. Women act as incarnations of the goddess in sometimes taking the life that they can carry, and in doing so, they enforce and control the particular segment of Nature's economy that their childbearing capacity represents. Indeed Nature commits abortion for reasons analogous to those held by unwilling mothers.

In treating human fetuses with a fine disregard for the presumed superiority of humans, Nature is less arbitrary and less sentimental. About 10% of pregnancies naturally miscarry, weeding out defective fetuses, conceptions that couldn't survive under ordinary circumstances. For some people, though—the rich or well-insured—circumstances are no longer ordinary. Now marginal pregnancies can be carried close to term and neonatology can do the rest. The high-tech medical care it takes to save these extraordinarily wanted and cared-for kids costs many thousands of dollars. These are fetuses that get treated like very important persons, but also as extensions of machines, delivered on the hospital's schedule and completed in mechanical wombs.

The existence of these technologies is blessing and confounding: I am personally acquainted with a few of these high-tech kids and can attest to the wonder and value that their lives have brought. It's uphill, and it's worth it, but it sure as hell is unnatural. Their parents and the whole apparatus of modern medicine strove hard to see that their potential to be was realized. Thanks to such mighty desire, the age of [technically assisted] "independent" viability has dropped below the age at which some embryos are aborted.

This raises hellishly difficult questions for potential parents and the medical personnel trying to serve them. It blurs what used to be clearer differences. These questions, I think, call for a wider, more thoughtful array of answers than for a uniform code that will fit every circumstance poorly.

Back in the days before the anti-choice backlash, the romanticization of the fetus, and the technology to make marginal fetuses viable, it was not uncommon to see illustrations taken from comparative embryology that showed human fetuses to have more in common with embryonic fish, frogs, chicks, and pigs than with, say, Miles Davis or Judy Chicago. Today that would be sacrilege, I guess. The point is that if you choose to look at sufficiently distant stages on the developmental continuum, the differences between fetuses and persons are self-evident. Eventual possession of little tiny feet notwithstanding, the fetus at first trimester (which is when the average abortion is done) is radically different from a newborn infant. The significance of that difference lies in exactly the same place that the question of the fetus's personhood does: the realm of human opinion.

It is a matter of opinion, of choice, of desire, and of timing. The weight of evidence of the personhood of the would-be mother is heavier than that of the evidence for fetal personhood. In spite of that, there are a lot of folks in the world who would make a virtue of overriding the evidence.

One of the most unnerving arguments against abortion nowadays is that there are thousands of people just aching to adopt babies (especially if those babies are white). These would-be parents are eager to treat unwanted fetuses as persons even if their unwilling mothers would rather not. This amounts to a proposal that women rent out their bodies for the purpose of gratifying someone else's desire—in this case, a desire for parenthood (which is not necessarily selfless—today's child often becomes a hapless accomplice to conspicuous consumption and parental narcissism). There are elements of prostitution in that arrangement (which practice ought to be anyone's right, but is condemned when the object is recreational, rather than procreational, sex).

Even in the lovingest of "new" adoptions, one wonders whether the biological mother's mentality—with maybe a tiny nagging awareness of serving as a luxury hotel with room service, difficult to maintain and staff—might not be hard on the fetus-person. Scandinavian studies showed that children born to women who were denied abortions developed significantly more social problems than wholly wanted children. This lends authority to the notion of significant communication between mother and fetus. It suggests that child abuse and exploitation can begin very early and that mental processes are as yet beyond the control of the state.

———

What *would* happen if we treated the fetus as a person? We would be denying an obvious biological difference between potentially and authentically viable life forms; we would once again be asserting that the highest and best use of a woman's life is childbearing; we would be negating the progress that has resulted from questioning that assertion; we would be removing yet another crucial responsibility from the realm of the personal and awarding it to society (and society has yet to prove itself to be a very great cherisher of the sanctity of persons); we would be making our culture even more disastrously anthropocentric than it already is.

At the mundanest and most actual level, if we treated the fetus as a person, things would not be so very much different than they are now. Hypocrisy would be even more rampant; women's lives would be even more difficult, and there would be a lot more litigation and law enforcement to do, which would employ burgeoning numbers of lawyers and hopeful legislators of morality.

———

A more interesting and inclusive moral question is: What if we treated the Earth and every living thing as a person? To people who regard life as an interrelated whole rather than as a hierarchy with man, and I do mean man, at the top of the heap, the answers are serious and imply a responsibility that transcends species identity. We would be forced to grow up and to acknowledge that living entails killing: requires choosing and mandates dying. We would continue to take life in order to live, just as we do now, but we might do so more reverently, conservatively, and personally.

So in trying to save the world, I would not pray for a miracle that would render humans desirous and the planet capable of supporting every human blastocyst that could possibly be developed. To do so would expand the cult of human personhood, and there are values far more important than that. I would pray for the human species to regain its sense of participation in nature, for recognition of the human lot of suffering, for the compassion and discernment to understand that the maximization of human numbers and lifespan is a primary factor in the simplification and reduction of the biosphere. I would pray for the restraint to act in accordance with those understandings.

Nulliparity and
A Cruel Hoax
Revisited

(1997)

A while back at my regular weekly women's meeting, I sat among friends. One woman, lacking child care, had brought her new baby daughter. While Mom ventilated the emotional strains she was experiencing as a single parent, baby Felicia captured every heart in the room. Most of the women could barely restrain themselves from snatching her out of the arms of whoever was cuddling her at the moment. It was a sweet, primal disturbance of our adult conversation. Then another woman, a tough-minded news hen and something of a jock, spoke of the pangs she felt putting her youngest child on the school bus for the first time and wept.

Clearly, mother love is a force of nature, easily trumping mere reason. Dave Brower used to say that you couldn't reason prejudice out of a person because it didn't get in that way. Reason is a pipsqueak, the melting tip of the iceberg of mentality. Which kind of makes me wonder why, back in 1969, I was so sure that I could and would get through my natural female life without becoming a mother.

I became a notorious non-mother when I shocked the media and my classmates at our graduation ceremonies with a commencement addressed titled "The Future is a Cruel Hoax." I declared that given the seriousness of all the eco-catastrophes then gaining momentum, "the most humane thing for me to do would be to have no children at all." An amazing amount of uproar ensued but my gesture manifestly didn't launch a mass anti-mass movement—not if all the baby-having going on around me, or the absence of overpopulation as a subject of concern in the public mind is any indication.

While I consider myself to be a staunch feminist, the largest system about which I can care is not womankind or humankind but Earth's evolutionary processes. Because it's axiomatic that wilderness preservation, restoration, and expansion are the minimum conditions necessary for this process to continue, my ultimate loyalty is to the wild.

Ecocentric, biocentric, animist, alone in a world of wounds—strange is the lot of those who chance into the deep-ecological mindset, who believe that "our community" means the ecosystem, watershed, bioregion, biome, continent, planet—all our relations; that every living thing is as important as any person; that they all could get along fine without *Homo sapiens* but not us without them. It's humbling and troubling; makes one feel like a grinch and superfluous all at once.

Population is, let's face it, a horrible issue. It's quantitative, parsing the richness and pathos of human life on Earth in incomprehensibly large numbers. It's an observable reality, but because exponential growth is not a sudden event, overpopulation remains somehow below the threshold of being perceived as catastrophic. As Garrett Hardin observed, "Nobody ever dies of overpopulation."

Here in northwest lower Michigan our pretty rural landscape—never mind the howling wilderness—is dying of overpopulation. Perhaps it's progress that nobody around here is in favor of just plain growth any more. They want the sustainable kind. I'm about the only person I ever hear wishing that people would quit having children. And because it really is an offensive thing to say, I do so only rarely.

In my community, baby-having and child-rearing automatically justify all manner of hyper-consumerism, from the use of disposable diapers to the acquisition of a family van, to trips to Disney World and a succession of pairs of $100 sneakers. In the utterly atomized nuclear family, "parenting" seems to have become a major job of work, for mothers, mostly, and therefore warrants such indulgence. Whereas among those unselfconscious, backward ecosystem peoples we hear that babies weren't the individual's or couple's property, privilege, or sole responsibility. There were fewer, happier, less fashionable babies (and slicks of baby poop on the cave floor). I have found that not having children is a great time-saver and an easy way to shrink one's ecological footprint. In conjunction with authorhood, a notoriously unremunerative calling, non-motherhood has kept my ecological footprint positively dainty.

In an interconnected world, the decision to bear a child isn't only a personal matter, nor does it pertain only to one's moment. Won't even the wanted, cared-for children feel betrayed to discover (assuming that such thoughts are still thinkable in the future) that previous generations ignored the problem of overpopulation and dodged the difficult choices in favor of a comfortable, conventional existence whose price included migratory songbirds, large mammals, old-growth forests and polar ice shelves?

I bite my tongue a lot. I don't want to risk alienating my friends, or nowadays their daughters, by arguing against their childbearing, except in the obliquest ways. Regardless of which birth it is, first, second, or third, I wind up congratulating new parents, especially mothers, warmly. At that point the horse is out of the barn. New parents have plenty of crap to deal with, even without a population bomber's disapproval, and children need and deserve to feel welcome once they're here.

As I push my grocery cart down supermarket aisles of sugar-frosted fiber puffs, overlit thoroughfares grid-locked with parents often rudely, and sometimes abusively attempting to appease or curb the advertising-inculcated desires of their TV-transmogrified kids, I find myself wishing that it were somehow possible to get my fellow Americans to be at least as thoughtful and caring about these children they've already had as they are about their cars.

In my youth I came across a women's magazine interview with illustrious non-mom Katherine Hepburn. In it Hepburn said she didn't think she could be as good as she wanted at being an actress and a mother both, so felt she had to choose between them. Fortunately for film fans, she went with acting. It struck me as eminently reasonable that one should assess oneself and one's society realistically and then make a considered decision as to the likeliest way to spend one's life.

Thus when women of my cohort and younger bewail the difficulty of combining motherhood and a career; or how hopeless it is to get their husbands (if said husbands are still around) to take on some responsibility for doing the wash or schlepping the kids around, I have to bite my tongue prit' near off. I'm sure that parenthood is exhausting. I agree totally that in contemporary circumstances the gender-based division of labor is grossly exploitative of women. But I have to wonder whether these women imagined that the revolution would be accomplished before the end of their pregnancy.

———

People refuse to believe the rules apply to us, that human beings are subject to biological constraints. The reasons for this exceptionalism are various—theological, ideological, technotopian. Me, I'm a Rules Girl. And, minus human exceptionalism, things are looking grim.

As the most hard-nosed population biologists have been patiently pointing out all along, if we do not address overpopulation by using birth control, Nature will deal with it by overriding death control. Given global climate change, sprawling megacities, declining nutrition, assaults on our immune systems, drug-resistant pathogens, and, with GATT, the prospects of no impediments to the worldwide movement of agricultural commodities and their hitchhiking pests, to say nothing of the possibility of rogue bugs bolting from germ warfare or genetic engineering labs, an awful lot of epidemics may be in store. The current opinion seems to be that death itself should be curable and, whenever it befalls, it's a tragedy. When the myth that modern medicine has conquered, or should be able to cure infectious disease is shattered, we will have a lot of philosophical maturing to do.

"Fear of individual death and grief," wrote Gregory Bateson, "propose that it would be 'good' to eliminate epidemic disease and only after 100 years of preventive medicine do we discover that the population is overgrown" ("Time Is Out of Joint" in

Mind and Nature: A Necessary Unity). These days, as forensic anthropology attempts to probe our deep past, some say that the growth of human population has steadily driven the series of technological changes: extinction of Pleistocene megafauna, thus hunting and gathering, then agriculture, and civilization, industrialization, and globalization—now approaching apogee. Thus, checking epidemic disease is only the most recent factor in the long, lurching history of the expansion of our species. However Bateson's insight that "fear of individual death and grief" are driving forces of our disproportion with the rest of life illuminates the core dilemma of overpopulation. Among individual human beings, birth brings joy and death brings sorrow. Forgoing children and suffering natural death will always be very tough to sell, given the abstract, almost absent nature of the rewards for such an ethic.

I've got a friend in her seventies who's dying of cancer. She's been relentlessly introspective, inquisitive, and iconoclastic for the decade of our friendship and is facing her demise right in character. When I asked her, "What is the meaning of life?", her answer was, more or less: it's no big deal. She intends no argument for living carelessly, but it's an interesting summation of a life of self-examination, spiritual exploration, artistic creativity, philanthropy, and humanism. Not nihilism, but liberation into a detached, non-anthopocentric relation to the cosmos. Fine for her, but what about those of us left to mourn her? It's going to take some pretty heavy philosophizing to get the human race to consciously check its will to love and will to live.

If a lot more women—say 90%—would follow my sterling example of nulliparity, it would unravel the biological family, seed-syllable of human culture and make for a wrenching, possibly disastrous discontinuity for our kind. Yet the need to contain, restrain, and minimize our species *vis-a-vis* more-than-human-nature is extreme. Earth's in a highly unnatural state of affairs. Can we be unnatural enough to regain our just proportion to all the rest of life? Which is the greater distortion of human essence: not to reproduce, or to live in a completely anthropogenic environment, every terrain dominated and depleted by the human species?

Deep down inside, population is nothing if not a women's issue. Personally, I wish that billions of women would just say no to motherhood and set up Amazon republics instead. All men have to do then is take their matters into their own hands. Of course it would be marvelous if ecocentric men would organize "snip-ins"—mass vasectomy festivals. To reinforce and reward this behavior, urologists could tattoo a beauty mark on the vasectomee's face above the beard line once he's flunked the sperm test. Kind of an antithesis to the semiotics of the wedding ring.

Once birth control and abortion are universally and freely available and the various pronatalist policies tucked away in the tax code have been abolished, but artfully, so that children don't wind up deprived as a result, propaganda might be the one acceptable means of civic action available to deal with overpopulation: an all-out attempt to change public opinion about reproductive behavior. And I'm not talking about a "stop at two" or even "one is plenty" campaign, but "Don't Do It!" There needs to be a steep decline in human numbers. Our last chance for it to be volitional rather than apocalyptic is for the vast majority of people now on earth not to reproduce.

The trouble with propaganda for non-parenthood is that it has tended to be tacky and materialistic, dissing children and gushing about all the fun you can have (read money to spend) if you're not buying magnetic alphabets for your refrigerator door. Economic calculus has yet to vanquish the drive for procreation. For just about everyone but the Amish, children are a major expense, non-contributors to the household economy. Still *Homo economicus* keeps on making babies. I would like to think that this means that our hearts are still flesh, even if everything else about us is bent by economism.

Of course, if the idea of persuading people not to reproduce is too heartless and objectionable, another way to attack the problem would be to promote, even insist on, BreathAirianism. BreathAirianism is drawing your sustenance from breath alone. Although to date its most prominent practitioners have been unmasked as fakers, not fakirs, given to gobbling candy bars off-camera, genuine BreathAirianism might be a way to dodge the birth control bullet. OK—no more gloomy talk about overpopulation. Have all the children you want, just nobody eat anything. Or go outside.

IS THE BODY
OBSOLETE?

(Response to a question from Whole Earth Review*)*

Is the body obsolete? Literally, this is a senseless question: senseless and sinister. It pisses me off because, if somebody in a lab is asking it, probably billions of dollars will be spent soon to answer in the self-fulfilling affirmative.

One could try to answer body part by body part: Is the nose obsolete? Has the asker lately savored the incomparable sweet perfume of an infant's scalp, the directive funk and tang of a lover's crotch, the fragrances of last year's half-decomposed beech and maple leaves intermingled with wild violets? Are hands, backs, tongues obsolete? Has the asker lately dug in a garden, ached from spading, clenched a fistful of freshly turned earth, or noticed that a vegetable's freshness is a tactile quality on the tongue more than a flavor, as in the ambrosia of potatoes uprooted moments before?

It's not the *body* that is obsolete: certain manmade conditions are pernicious and must be reversed. The correct response to finding the mineshaft canary dead on the bottom of the cage is not to build a canary that can't be asphyxiated; the correct response is to return to the surface immediately and possibly even to seal up the mine. It takes a strange and alien attitude to regard a living thing as unsatisfactory. They say that torturers and playground killers are characterized by a blind spot: an inability to imagine that their victims' insides are real. Does the concept of bodily obsolescence bespeak a similar dispassion?

To declare the body obsolete is to settle for less, not to acquire more. If we want a biological planet (although not everyone does, apparently), then we must heed our own corporeal biology. It befalls us to be the chief canaries-in-the-mineshaft. A robotics that will allow us, by proxy, to splash blithely in toxic or radioactive substances will sever a feedback loop and help to diminish our perfectly sensible fear of these excrescences.

To me, bodies, whether my own or those belonging to others, are impossibly wonderful, entirely elegant, even— strangely—in infirmity and old age. The body of any organism is de facto perfect from its first moment to its last, a particular expression of the total and local biology, most meaningful in context, interpenetrating with the world that brought it forth.

That the human body is not invincible or omnipotent seems highly desirable to me. Human consciousness is obviously too immature to function morally under either condition, which is why the roboteers who would presume to direct evolution on the basis of mere ego are dangerous men.

When you hear all the techno talk about "Us" directing evolution, bear in mind that "Us" turns out to be mostly tall white guys with good teeth, the same crowd that for centuries has been dependent on the physical labor of wives, slaves, children, and laborers: proxy bodies regarded as more dispensable than those of the father classes. A lot of overreaching was accomplished by these suffering proxies:

looms tended, coal mined, rails laid, towers flung up, and cannon fodder borne.

Behind every anger is a fear. Behind my anger with the radical materialism of these hypothesizers is the fear that, once posed, the question of the body's obsolescence will be seen as relative. I think it would be safer to see it as absolute. Either the body is obsolete or the body is holy. Can't have it that the body is obsolete in certain cases and holy in others.

Our best wisdom is founded in our flesh. It is generated by living tissue. Through embodiment, the Buddha was brought to awareness and Christ's incarnation sealed the saving sacrifice. And long before those enlightened self-abnegators, long before it became a man's world, we lived upon Mother Earth. The primal memory and original fact of the human species is the awesome life-giving power of womb, vulva, and breasts: the sudden deluge of amniotic fluid followed shortly by the emergence of an entirely new Thou.

"Is the body obsolete?" is a jaded question best answered by well-placed lightning bolts, if wet chicks cracking their way out of eggs or the gnarled prolific hands of an aging Monet or Renoir don't do the trick.

RELIGIOUS EXPERIENCE

SPIRITUAL SWIMMING

(1983)

For years I swam in pools for exercise, but it felt like a chore. I did it resignedly, in a mundane frame of mind. Recently, though, I began swimming in the open water of Aquatic Park on San Francisco Bay and, also recently, began trying to live according to some general spiritual principles; to strive for conscious contact with god. (My current working concept of god is the benevolence of anything and everything, including, but unbounded by my self. I've got no dogma, and my discipline is strictly homegrown.) Now, early in the morning, when I go down to the water's edge, I make my devotions. I swim toward god every day.

One might ask whether it's not a tad sacrilegious or, worse yet, self-serving to elevate a fitness regime to the realm of the spiritual. Could be. But my spiritual advisors tell me that mindfulness and gratitude are qualities to cultivate all day. Elements of character don't just grow in pews or on zafus. So—spiritual swimming is a part of an ongoing, life-enhancing effort.

What makes swimming especially conducive to this cobbled-together practice is its particular nature. It's not earthbound and therefore not a pounding, punishing activity like some

I could name. For a little while each day, it frees me from the restraints of gravity. Like yoga, it takes stretching and extending. A quiet invigoration comes from doing it at an effortless pace. And, like yoga, swimming requires attention to the breath.

When I'm doing it right, focusing on a spot straight ahead like a ballet dancer holding a spin, face submerged just up to my eyebrows, the world consists only of translucent gray-green water broken by the flash of an arm trailing sparkling drops, glimpsed mid-stroke, mid-breath. It's simplifying and quieting to focus on just these things and to experience that fierce cold water on bare skin.

The crawl establishes a meditative rhythm. While I swim, I think prayers or mantras (and worry and recriminate and plan and gossip; the balance of one mental activity against the others indicates the spiritual condition of the moment). There's an endless way to go in my quest and through the water. The athletic wisdom of pushing oneself a little further than before has wide applications for the development of spirit as well as flesh. The prayers are formal. And deeds can serve as prayers. As one spiritual advisor suggested: "Consecrate your struggles to god."

On winter mornings, when great storms out at sea manifest locally as four- or five-foot waves breaking between the piers on our sheltered beach, it's frightening. I'm scared of being caught up and dashed on the sand. Surrender is the only way in. Count the waves as they come in series. Sidle in as the least wave is breaking. My swimming advisors say, "If you get caught, duck under it. Let the crest and its friction pass over."

Once beyond the surf I'm as safe as I'll ever be, and the mantra kicks in: I love god and god loves me. Four strokes. God's love is in the physics which buoy me up and in the intricacy of my whole being, moving me to and through the water, bearing my mind and all its yearnings: everything that is and lives is holy, even me.

Part of this quotidian spirituality is trying to grasp the lesson of the day's events as parable, and I like my teaching stark. During the round of the year, the water temperature fluctuates by about fifteen degrees, and each degree makes a difference, decides our relationship, determines my actions. I'm only a particle of reality. In January, when the water goes brown with soil from upstream and becomes a soup of tule rushes brought down by bad weather, that truth becomes vivid. The water screams and bites at the skin and feels great! Shuddering into it is a worthwhile ordeal. The time and distance I can make depend on the cold. To spend fifteen minutes doing a quarter of a mile takes eternity.

While I swam one morning this past winter, exhaling hot breath with every stroke, it hit me that I'm just a transformer, a way station for matter's transit toward ultimate disorder. Food, organized life of the land, was fueling my life's motion. Bitter cold drew that energy clean out through my metabolism. The warmth just dissipated aimlessly into the frisking bay. For that moment I knew my job was simply to be in passage, and that the way to do it well was to be at one with change.

Learning that strength will come in proportion to one's embrace of elemental reality is the great work. It's the effort to love god and to build faith in god's love. Faith and trust can't be contingent on getting one's own way. Quite the contrary.

Swimming is easiest when the water is calm and warmer. You feel you could go on forever, or take a little nap, and that's a blessing to count. But when it's choppy, there's no acting like a well-oiled machine. The surface is lively. To keep time with its syncopations, you have to get loose, like a sea animal. God is a challenge to meet changing situations with grace. God is a shaping grace.

Acceptance like that comes on good days. Other times, the task is wrestling the fears and crazy mental noises which begin to roar in a lonely mind. This silly moment, for instance, taught me something: I was about as far from shore as I ever go, spaced out with the chill and exertion. (Lowgrade hypothermia mildly alters the mental state and can render you susceptible to terrors and revelations.) I noticed a swell building under me. I imagined its being a tsunami (one archetype of devastating transformation). Reflexively, I thought, "I must *do* something!" That is a long way from surrender. Fantastic though the problem was, a message finally got through: I could do nothing. Those modest swells rose and fell and broke harmlessly on the shore. Floating on them I learned my powerlessness and then swam on.

ST. HERPES

Ages ago, I got caught up in a zeitgeisty dinner conversation with a group of artists and writers. The leading question was: were people still doing it on the first date or were they waiting awhile? The consensus was that our contemporaries had become more conservative about sexual encounters. Like a baleful beggar at a feast, I commented that the new restraint probably had something to do with sexually-transmitted disease.

The conversation turned immediately to herpes: the late twentieth century's most celebrated non-life-threatening disease; the new leprosy. My interest—and my discomfort—were acute, because I have herpes. (This evening was quite a lesson in finding out what people really think about your kind when they don't know what kind you are.) By now, everyone has heard all the herpes jokes: What's the difference between love and herpes? Herpes lasts forever. What's the cure for herpes? A leap from the bridge. Depending on who you read and who you're with and what vantage point you take, celibacy or death might seem your only options.

Over the years that I've had herpes, my feelings have traveled from isolation and despair to resignation and gratitude, with a few shades of indifference between. When my spiritual condition is fair and my mystic acceptance quotient high,

I refer to my complaint as Saint Herpes. This ubiquitous virus has protected me from a lot of amorous pratfalls that my waking ego was willing to take. Not that I believe there's anything wrong with falling in love and doing all the beautiful and dumb things that Eros commands, but the wear and tear can smart. If this is a masterpiece of rationalization, it seems preferable to pathetic self-pity, which also afflicts this herpes carrier from time to time. When you have a lemon, make lemonade. At other times I contemplate the attractions of chastity, a practice quite different from the one that brought me to this pass.

Knowingly, voluntarily, I exposed myself to the disease a few years before the late-seventies hullabaloo, when herpes became a social fetish and conferred instant pariahdom. So, because I did want to have sex with the person in question (who had done the decent thing and informed me of the risk and was willing to practice prophylaxis), I contracted herpes. Less was known about it back then than now, and far less was said. Fewer people had it, but the epidemic was building up steam. That was a time when I slept with people for love and fun, for the sake of sleeping with them; slept with them for the joy of sex itself, or for lack of enough gumption to say no. I was in my twenties, it was the seventies, and abundant sexual encounter seemed like my birthright.

And today, as far as I'm concerned, that is still a fair belief. You make love or you don't, you connect with your partner or you don't. Sex is special; sex has overarching power; and sex is free. Every good thing has drawbacks, though, and the enthusiastic exercise of sexual freedom is not exempt. The fact is that the more partners you have, the greater the possibility that you will contract a sexually-transmitted disease, and/or get your heart broken, and/or learn an incredible amount of good and intimate truth, the kind of truth that is revealed only in sex—in your lovemaking. No blame here. Sexually-transmitted disease is a possible consequence, not a punishment. Living is full of consequences, and it binds us to change. So this is no repentant sinner talking, no moral majoritarian argument against free love, no case for herpes as an agent of divine retribution for sexual pleasure. Mindful or mindless, caring or careless: this most natural of acts is never without some meaning.

When the disease actually showed up, I was dismayed, but only briefly. I had no way of knowing then that herpes would some day become a factor in my changing from one kind of person into another.

What makes herpes such an existential curse, no matter how unobtrusive it may be in a given individual, is its communicability. Even in the absence of symptoms, there is no guarantee that it won't be transmitted. Such transmission is highly unlikely, and condoms can render it even less likely, but there can be no certainty. Consequently the new herpes etiquette demands that you always inform your prospective partner that you have herpes, leaving the prospect to conjure with the risks. The psychological discomfort of those moments, the strenuous effort to play fair and tell the truth and care for the other while detaching from hopes and expectations, far exceeds any physical pain that genital sores can deliver.

For years, my herpes and I were safely out of public commerce in the confines of sequential monogamy with two fellow herpes victims. (The fact that in each case we had the disease in common was coincidence—it wasn't something we even thought to discuss.) It posed no big issue in either relationship. In the first, herpes outbreaks would mandate periods of abstinence; that was a strain, but not the one that eventually sundered the tie. Occasionally the virus precluded the possibility of helping reconcile differences erotically, achieving the closure that lovemaking can bring. Herpes between lovers is not nothing, but the decision to risk it is behind you, and the experience of the disease has its container of intimacy: the advantage of herpetic monogamy is that you can get past the preoccupation with the disease and into knowing and delighting in the presence of another human as person, not vector. Within a relationship or out, the worst thing about herpes is that there are times when you want to make love but can't. Desire goes ungratified (as it often does in the uninfected).

Relationships tend to pass. It falls to most of us sooner or later to leave the safety of monogamy for the renewed terrors of dating, if only for a short time on the rebound. It was when the aforementioned monogamous relationship ended that herpes emerged to add to those terrors and to start reordering my life. On my mad dash to the first available bed, my herpes, long dormant, kicked up like nemesis: a bit of protein with perfect timing. I began to feel vague inklings of herpes (the "prodrome") hours before the likelihood of lovemaking became clear. Uncertain moral actor that I am, I almost didn't tell the man in question. I wasn't sure whether those vague twinges could infect anyone else (medical opinion is divided on that point). Besides, I was sex-crazed by the post-parting fear that I would never know love again. Fortunately, superego whupped libido and made me speak up. Instead of winning the man's ardor with my late-breaking honesty, I was compelled to cope with his honest aversion to contracting herpes. We cried about it together. He hoped we could be friends; if mere friends would be too much of a strain on my makeup, well then, so be it in sadness. (Thus the disease effects a role reversal: the gentleman pleading, "Can't we just be friends?") I had nothing to lose by this, so we became friends. Friends with a constant taint of anger and resentment on my part. Whatever else that turning point in the relationship signaled, the nice fellow's sensible withstanding of my allures registered as rejection. In a trice I was stripped of my accustomed power of sexual sorcery. It was a heavy blow to the pride.

Alas, having herpes does tend to make one undesirable (except, perhaps, to a fellow sufferer). After reckoning with this new reality, I had to begin to place my faith in my intrinsic worth, quite apart from gender and relationship. Identity can deepen in crisis. Part of the herpes educational program was discovering that there is life after rejection.

For all the dread that herpes occasions, the physical experience of it has been uneventful, not painful; mine never has been the nightmare version of the disease that has struck fear into the free hearts of swinging singles. If it weren't for other people and desire, herpes would be less annoying than acne. My experience, apparently, is typical. Painful recurrences are exceptional. Herpes is a major health hazard only to neonates and the immunosuppressed.

I have learned that such a disease creates a personal relationship of its own. Herpes is different things to different people because, I guess, of the way viruses operate. A virus infection is a form of possession. Viruses are invisible colonizers, subcellular entities with an elegant survival strategy of invasion and takeover. These fiendish particles of nucleic acid are everywhere. They pass through filters, maybe even through condoms. Once the herpes virus comes on board, it's a lifetime companion, albeit one that may retire quietly into the nervous system after its initial appearance, never to be heard from again. Or it may be like my friend St. Herpes, a bawdy crazy jester that drops down for a chat every time I feel horny, or whenever there's a man around, just to see whether I'm paying attention.

Herpes is thought-provoking and thought-provoked. I am convinced that herpes is so mutable because its stronghold is in the human nervous system. Stress activates herpes and herpes activates stress. Trying not to think about it is like trying not to think about your tongue. Even people without herpes know that what goes on in your mind can arouse distinct, uncontrollable feelings in the genitals. So it is with herpes. It becomes an integral part of you.

Herpes has shown me how selfish I can be. It has engendered a sexual greed so intense that I could just about convince myself that the symptoms I felt weren't real and so get it on without delivering the warning and risking the rejection. The subtlest symptoms, the "prodrome," may or may not lead to an outbreak of sores. The sores are highly contagious, the prodrome, just possibly. Responsibility would seem to demand forthright presentation of the worst-case scenario. However, responsibility gets harder to come by when you're beginning to wonder whether you'll ever get laid again. The prodrome, sez I to myself, could conceivably be hypochondria. Even a medical lab couldn't say for 100 percent sure whether the virus was actually "shedding." The only way to achieve certainty in the situation would be to refrain from intercourse. But sex had become a fixation for me, so I did the despicable a couple of times, sleeping with men during what might have been a prodrome without reading them their rights. Maybe that these partners had been apprised of herpes as a party to our affairs makes this more forgivable.

In neither instance was the disease contracted, but that yearned-after sexual connection was, for me, made impossible by guilt. One romance was demolished by herpesphobia when my partner confronted me, having developed some symptoms (but not of herpes, to my infinite relief). His imagination was worse afflicted than his body, and he was angry with me. Then I was angry with him. He'd been warned—somewhat—and the risk was now his responsibility. Another lover got some mysterious sores in his mouth and coolly decided that he was not falling in love with me after all. He informed me, by mail, that, pleasant as our keeping company was, the anxiety was not worth it. Only after my hurt and dismay at such partings subsided could I ask would—could—this love affair have turned out differently without herpes as a factor? How important was it, actually? Or was St. Herpes watching over me?

Guilt thrives in the moist atmosphere of sex, and anguish grows where there is a void of shared beliefs. Herpes, I am concluding, is none other than a situation-heightener, throwing into sharp relief the human stuff I'm made of—I prove to be, among other things, hungry, needy, frightened, compromising, and sly.

As my sexual encounters became problematic and rare, I began to crave sex in the abstract, a disembodied sex, as commodity: sex for its own sake; but now it wasn't so easy to come by. Distant contemplation bred molecular appreciation of the sex drive. I began to feel it in my bones, in the cells of my body, in my chromosomes and midbrain. I became the egg's immensity of wait and spherical receptiveness. I wanted the enlivening presence of millions of madly traveling, take-a-chance, go-for-it potentiators. I missed the scramble of sex, the wonder of finding one body inside another, with currents of life coursing past boundaries of self. I began to perceive sex as a raw force by which our clever, thoughtfully-crafted selves are driven like robot cars to sudden collision, with Mother Nature cackling at the evolutionary controls. It takes the power and lure of sex to breach persona and ego and make monkeys out of the experts on healthy, sensible love. This transcendent understanding of sex was a gift from St. Herpes. Away from sex, I pondered its holiness, the perfection of surrender it offers, and earthly version of the perfection that lies beyond opposites and distinctions. Irony compounded irony. If my body wasn't getting any, my mind surely was.

A little obsession goes a long way, though, and it became a mistake to think of sex outside its personal context. As a result of scarcity, I developed what my friend Sarah diagnosed as "penis-vision," a tendency to regard all men, even and especially slick and greasy characters whizzing past on the freeway in their big cars, as potential lovers.

The catch is that adapting to herpes demands a temperate, considered sexual ethic from a person who wouldn't have contracted herpes in the first place if she had had a temperate, considered sexual ethic. Herpes and other of life's learning experiences have convinced me that being hit with the cream pie of fate is a clear sign that the lesson was necessary. I don't believe in abstinence or privation as goods in and of themselves, but learning to live with them ultimately has been beneficial to me.

The problem is, as Anne Herbert wryly observed, "It's hard to live ultimately." Philosophizing was swell, but it didn't alter the reality that herpes murders romance. Sure, it's possible to abuse romance, just like a drug. Nevertheless, even when the killing of a romance may be a mercy, euthanasia is still a death. Blind, flying, spontaneous leaps into the sack are a thing of the past. Every tentative embrace calls for a Miranda reading: You have the right to remain silent. You have the right to consult a physician. I have herpes. Here's some free medical literature. Can you read it in that candlelight?

It's just plain heartbreaking to have to wrestle the cherubs of infatuation down to the plane of intellect. Having to consider, at the very beginning of an affair, whether that affair is likely to be worth catching an incurable nuisance of a disease is miserable. Yet one must make that impossible decision before the sweet poetic fulfillment of mutual fascination can be found in a lover's arms. It takes an insanely romantic temperament to make such an affirmative imaginative leap. And because of the fragility of relationships, most of today's possible partners are battle-scarred veterans, so emotionally flayed and slow-healing that they doubt that any good thing between the sexes is either possible or worth daring, herpes or no. The thought of the disease just compounds the dread of emotional pain. I share that dread. Yet I know that the elaborate, unconscious craziness that proliferates in avoidance is worse than pain itself, more confusing. Pain has the decency to ask a simple question. Fear just pretends to have an answer.

So herpes becomes an agent of natural selection, weeding the faint hearts out of the available universe of lovers. It gives ample reason for turning back, a handy focus for all the doubts and hesitations that smite the smitten. For the man with a deeper interest (or his own case of herpes), the disease provides a point to ponder. Even in the most amorous affair, there's a lot more communicating and negotiating per day than there is lovemaking, so partners might as well broach and resolve a complicated situation beforehand, just for practice.

Because of its habitation in the nerves and, by extension, the emotions, herpes has a quality of being compensatory, or moral, or asked-for. It's difficult (and not altogether truthful) to objectify and isolate the disease as not-me. Yet it's necessary to talk about it rationally without being defined by it, in spite of the fact that the disease reflects one's inner life. Herpes shows me how I treat myself, how I enjoy the adrenal glamours of stress. When I'm hectically caught up in my own importance, I'm vulnerable to an outbreak. So I have to be humble enough to sleep. There are limits to what I can or will do to fend with my herpes, but it always brings me back into awareness of my body—it makes me mindful of what I'm consuming, it notifies me when I'm defying my natural limits, and it actualizes the penalties of living in the head, or paying exclusive attention to the chatter of waking consciousness. Meditation, they say, may alleviate herpes.

In the midst of all the high seriousness of the herpes uproar, it's been commented that this is mainly a worry of the affluent, the kind of physical complaint that only becomes noticeable in a context of prosperity, good health, and sexual freedom. Although accepting the conditions the disease imposes has taken a long deep while in my own life, I know that herpes is also trivia, just a twist of fate that sets things in motion.

I would argue, then, that having herpes is having a guardian angel, one who does the wise thing for you that desire won't let you see. Having herpes is being a princess in a tower in the midst of a forest of thorns, a princess awaiting a prince who'll risk his flesh for love. Sometimes, enervated by hot, dashed hopes, I come to my senses and awaken alone, relieved to be unencumbered. I light a mental thank-offering to St. Herpes, the Exasperating.

The resultant paring away of my conquests and love affairs has been a blessing in heavy disguise. In the sizable interludes between brief melodramas, I've come to revel in my solitude. Herpes has taught me to live by myself. Partly from resignation and partly from the pleasure of my own company, I find myself coming to cherish the small solid pleasures of my daily round, with no mate to chagrin my morning prayers.

The big project before all of us always is self-knowing and self-forgetting, opening the higher lines of communication. Herpes is the dragon at the gate, securing my privacy and mocking my lust. It turns out that we need each other, the virus and I. I've thought about it long and hard, and I've concluded that herpes is life's way to make a point about sex with honor. I don't want my body writing checks that my soul can't cash.

Earth And Her Discontents

EXCERPTS FROM IN PRAISE OF NATURE

(1990)

EARTH

It's all too easy, in the dailiness of existence, to begin to take life for granted. And yet that there is life, and that life comes together in so many forms from so many elements and smaller forms, is never less than astounding.

"A mouse is miracle enough to stagger sextillions of infidels," wrote Walt Whitman. Life on Earth is the magic you can watch minute by minute. Every single cell is a wonder, a cooperative community of subcellular bodies that have come together in the destiny of being a larger metabolic unit. How many millions of cells, how different one to the next, differentiated from one tissue to the next, how many of those miracles compose the unity that is a trembling field mouse? In one human body, we know, there are a hundred times more cells than there are stars in the galaxy.

For oxygen to be present in Earth's atmosphere in an amount useful to air-breathing life-forms took eons. The development of soil and its fostering of millions of species required billions of years. Vascular plants, among them the grasses that make all flesh, are a sophisticated development. Woody stems, root hairs, capillary action drawing minerals and water up toward leaves for transformation through photosynthesis—flowers and seeds are inventive measures for life to have taken in its ongoing play. We all live by that greenness: photosynthesis is the limiting factor for almost all life on Earth. Even though we may procure it at a supermarket, our sustenance is entirely from nature, and in nature it cycles through. Eat or be eaten is the name of the game. There's no sustenance without something's dying, the dying in parts of perennial plants, the dying of prey in the jaws of a predator, the dying of predators and their reduction by the actions of myriad soil makers. Death drives reproduction, the paramount force in evolution. Transmission of genetic information is the aim of each being, transmission of characteristics and admitting to the possibility of change—gills to lungs, limbs to fins.

We don't know how many different kinds of life-forms there are. Estimates of the numbers of species range from five to thirty million. Because so many are as yet unclassified, microscopic, or obscure, nobody is quite sure how many there are, actually. However many there are, habitat destruction—the clearing of tropical forests, the logging of the ancient forests of the Pacific Northwest (and the upper Midwest), the dredging and filling of wetlands, the smothering of coral reefs with effluents from development onshore—is quickening the pace of extinction in our time. Epochs of extinction have occurred before in Earth's history. The late Permian extinction removed half of the families of marine life, and the greatest mass extinction to date took place in the late Cretaceous period and wiped out most of the evolutionary lineages of the dinosaurs. The human species is causing a mass extinction to top that, an evolutionary event the like of which hasn't been felt on Earth in sixty-five million years, taking an immense toll in biotic diversity. It's a dying on the inordinate scale, but we can take action to stop it, once we get, in our hearts and our bones, that our fate as a species is bound up with that of every other creature. We live one life.

Our early ancestors understood themselves to be as much a part of the Earth as anything else in their world, and felt that everything was alive and invested with power and mystery. A sense of the Earth as a great goddess, a bringer-forth of life and an enfolder of the dead, seems to have been widespread in Paleolithic and Neolithic times. Creation myths envision the Earth as a being—sometimes as a great animal. Native Americans saw their part of the Earth to be Turtle Island. They had a sense of the Earth itself being vital and long-living, as turtles are, tenaciously enduring. A turtle's heart can beat for a long, long time.

Until well into the Middle Ages, even most Europeans believed that Earth was, in a sense, alive. The Enlightenment, with its clockwork sense of matter, dismissed that understanding from civilization until the late twentieth century, when scientists James Lovelock and Lynn Margulis announced their Gaia hypothesis that together the planet, its life-forms, and its atmosphere are interacting and mutually creating, and have some of the properties of living tissue; that the Earth is like an organism. Thus, the Earth is not just a molten-centered ball of rock on whose surface life fortuitously happened to evolve. Rather Gaia (an archaic name for the Earth mother) lives.

Having access to great bodies of relatively new and rapidly developing scientific understanding is perhaps a blessing of our time. Geology, for instance, which emerged in the late eighteenth century, challenged the Judeo-Christian idea of creation by asserting that past geological changes were brought about by the same causes presently working on the Earth's surface, and by stressing the near-infinite slowness of the process of Earth-shaping change. Physical evidence showed that the Earth's surface was being unmade and remade, that continents were worn down and mountain ranges thrust up over the eons of time.

––––––

The imagination it takes to grasp the spans of time it took for the planet's surface to cool down, and to slow down, and for there to begin to be a less cataclysmic shaping of the surface than by meteor bombardment; and to picture the dance of the plates bearing along the continents and, with them, the arrays of species that would diverge and change once they were separated by new oceans and landforms, as the forests of Asia and the forests of North America did, is wilder than the imagination it takes to believe that it was all made in six days.

Earth seems ancient; yet there was a time when Earth did not exist, and then a long time, when it was inhospitable to life; then billions of years when Earth was amenable only to microbial life, whose presence over these great spans of time transformed Earth's atmosphere and made it possible for yet more complex forms of life, like nucleated cells and multicelled plants and animals, to evolve.

Strange to think about the dynamism of the face of the Earth, its continual transformation, and of the profusion and diversity of life taking hold everywhere it can. No two places in nature are identical; every inch of living Earth differs from every other, and in ways that may be fateful for their inhabitants. A particular forest—maple-beech or tropical rain—cannot just spring up anywhere. Ecosystems result from different combinations of soil, slope, elevation, precipitation, and proximity to pole or equator. The living creatures inhabiting ecosystems may be indigenous rarities— fish found only in a few desert hot springs, unable to survive anywhere else, or a salamander wanting to dwell in sluggish peace under a moist rock by an Olympic forest stream whose riffles are kept at just the right velocity by the impediment of occasional fallen giant firs. They may be more common and adaptable—like coyotes and opossums, two species that are currently expanding their ranges. Whichever the case, each species has a unique lifeway and an essential relationship to the health of the whole system, a role to perform.

Humanity is flatly incapable of replicating anything so complex, fortuitous, and subtle as an ecosystem. Ecosystems are the Earth's way of maintaining a dynamic equilibrium among groups of organisms in particular locales: not unchanging, but diverse enough to be able to integrate the changes taking place over evolutionary time. The more diverse the ecosystem, the greater its resilience.

Diversity is a measure of the numbers of species present. Thus species extinctions simplify, and destabilize, whole ecosystems. Because each ecosystem is different, quite place-specific, no formula counsels how far we can push the forgiveness of ecosystems in our desire to exploit the resources they provide or that underlie them.

As a tool-using, cosmopolitan omnivore, the human animal perhaps finds it difficult to understand how inseparable some creatures are from their ecological niches, how finely attuned to specific temperature ranges, nesting spaces, diet, and territorial needs. Change one too many of these critical variables and a species vanishes. Even before the last member of a species vanishes, its numbers may decline below a level at which there's sufficient genetic variation to allow its evolution to continue. Even though it may be possible to preserve specimens of endangered species in zoos, some qualities, real and ineffable, vanish in captivity. As David Brower commented on the decision to resort to a captive breeding program for condors, "A condor in a zoo may feel a breeze, but it will never know the wind."

It may be that we will be able to look back at the last half of the twentieth century as an extraordinary turning point in Earth's history. These were decades during which the human species became, and then became aware of being, a geophysical force on the planet. We have effected atmospheric change nearly as dramatic as that made by the cyanobacteria, which took two billion years to produce enough oxygen for a whole biosphere of life to breathe. Through deforestation and subsequent desertification, we have had a geologic impact more widespread than any chain of volcanoes erupting or plates colliding.

We haven't fully comprehended the immensity of this power, and civilization seems blind to the absolute necessity to learn the language of evolution—spoken by natural history—if there is to be a positive outcome of human history. For many people, the nearest opportunity to learn a little natural history may be in a park or a vacant lot. And while there is wonder everywhere there is life, it is critical to learn the difference between the hodgepodge of native and exotic species that may be found in a park or vacant lot and the coherence of relatively intact ecosystems. Basic ecological literacy would demand that everyone have some sense of the difference between ancient forests and second- and third-growth woodlands, because life hangs in the balance of knowing those details.

Besides, the experience of learning the details, which means going outdoors, is what can hearten us to segue from the twentieth to the twenty-first century in a lifesome way. As the winter solstice approached, snows in great variety fell almost every December day where I live, with deep white quilts laying a silence on the Earth. The soil might not have been frozen then, but the cold suggested a stillness in Earth's pores, where insects and microbes dwell. The pine boughs drooped with the weight of snow, and occasionally a red squirrel's hectic travel from branch to branch would dislodge a powdery clump and send it plopping down with a surprisingly audible impact.

That summertime sense of life's teeming was all but absent now. Cottontail tracks across the drive disclosed their presence, and a mid-December sight of two foxes, ambling and snuffling and scent-marking around the margins of the meadow, was a rare blessing. Conifer needles glimpsed through snow were the only witness to photosynthesis, and mammal footprints the only signs of wildlife. Otherwise, Earth's fertility, diversity, and activity all seemed hidden—and becalmed.

Except for the chickadees and nuthatches. Now these little birds are exceedingly commonplace, and you don't have to be Saint Francis to get close to them, especially if you are willing to top up the bird feeder regularly. Which is what I was doing, all bundled up behind my house one morning, wonderstruck by the whirring fervor of wings around me in all the hush. The longer I stood watching and listening, the more black caps and white cheeks I noticed glinting among the pine branches, perching and waiting politely for the chance to land at the feeder and carry off a black sunflower seed. To sustain their fast darting lives in the cold, they must feed constantly, stoking the tiny metabolic furnace with industry all out of proportion to their size.

Thus in the dead of winter I found myself surrounded by life and the white loveliness of the December landscape. Hardly a wilderness experience out back, and a monochrome-minimalist antithesis to the immense green riches of biodiversity imperiled in the tropical rain forests. Nonetheless, all that hungry activity was a powerful spate of Earth magic.

The chickadees' chattery insistent voicing of life's chant, it seemed to me, merited no less a response than the oath of the bodhisattva: "Sentient beings are numberless—I vow to save them."

AIR

Take a deep breath. Go ahead. Inhale till you feel your diaphragm move. What do you feel? Is it the blessed breath of life? *Prana* is what the yogis call it—breath energy. The act of breathing spans the conscious and unconscious, voluntary and involuntary body functions. Hence, focus on the breath is an ageless meditative practice. Rightly understood, every breath you draw is a gift from the universe. We are air-breathing animals. To live we need oxygen. In just the right amount, Earth's atmosphere provides the oxygen molecules to help us metabolize our sustenance—and the inspiration that animates us.

The air seems almost like nothing. It is the membrane through which we perceive, and by which we are protected from, the sparkling vacuum of the cosmos. The air is the enveloping atmosphere of planet Earth, inviting our sight to seek the heavens. Sometimes (or seemingly always, depending on where you live) the air bears countless tons of water, as cloud panoramas or as just an expanse of overcast (when as far as one can see is up to the ceiling). Thanks to the jet age, many of us have had the experience of looking at clouds from both sides now. We've passed through them and flown over them and cringed as pilots threaded their ways among them. Many more of us have watched the skies, with the realms of clouds, just for pleasure. Down to Earth, on a bright day, you may look out to a clear blue vastness, all illumined by the sun, simile of an empty mind. And at night, if you are fortunate enough to live in a region where ambient light doesn't interfere, you can gaze up at the stars, knowing that there are billions of them, and possibly millions of solar systems, some that might even be hospitable to life.

Because we humans are always looking for guidance (or justification), usually in the form of a story, it is perennially human to try to read the metaphors in the night sky, to tell the myths associated with the constellations, and to seek the governance of the zodiac, or of fateful shooting stars, or comets as portents of millennial change.

Migratory birds also consult the sky for guidance, if not of the metaphysical sort. They navigate by star patterns and the position of the sun on their journeys the length of a hemisphere, or across the trackless seas. Before we ventured up in balloons or 747s, the sky was largely reserved for the stunning variety of species of birds. However common, a bird is always a bit of a wonder. Even the neighborhood birds are a delight and an amusement. One late autumn afternoon in the course of writing this essay, I watched an impromptu convention of about a dozen crows taking place in a taller-than-average tree out back in the lately bare woods. These crows were a lively crowd; they seemed a little precariously perched, bobbing in the wind as the branches swayed in the chilly breeze. I love the way crows seem to drift, float, and quiver in midair. A couple of them will slope off, casually doing a loop-de-loop together for no other purpose than fun and sociability; then others lift off and alight, taking a quick look-see around the treetops. The why of it is a happy mystery.

As the province of flying things, the air was, until only very recently, a realm beyond our powers. Now *we* claim to have conquered the sky—flight paths have become to us what sea-lanes were to our great-grandparents. Not only have airplanes become a commonplace means of transportation in the late twentieth century, in space capsules we have escaped Earth's ambit and traveled far enough away to have a look back at the home planet and take snapshots of Earth as Gaian mandala.

Earth's enveloping atmosphere is the caldron of planet-girdling currents of wind, humidity, and temperature, the arena of tornadoes, blizzards, and drought. The weather is a mix of water and air driven by fire (in the form of solar radiation). Impeded and rerouted by Earth's landforms, given its dynamic by temperature differentials and the Earth's rotation, weather governs our lives. Over evolutionary time, each living thing has developed a specific relationship to the weather and climate. Most organisms are not so cosmopolitan as we are. Plants, being unable to flee, are particularly limited in their temperature ranges. Animals, being more mobile, are less limited by climate, except those creatures that are closely coevolved with a single plant species or association. And *Homo sapiens,* who can build shelter and make clothing, has extended her range throughout the planet. Because of the rootedness of plants, dramatic or sudden changes in the weather can be devastating, ecologically, to wild plant communities. Global warming, brought about by a man-made increase in the atmosphere's "greenhouse" gases, is exactly such a change. Cosmopolitan and footloose though we may be, because our civilization depends on agriculture and a relative handful of cultivated plants, also entirely subject to climate, variations in weather hold the potential to change history.

Weather is made of many forces. Among these forces, humans have come to know the winds as intimately as anything unpredictable and invisible can be known. We have been influenced by the winds, have blessed and cursed them: hot winds that make people do crazy things, winds that create conditions for wildfires or set the undertone of certain seasons, have names—chinook, simoom, samiel; foehn; khamsin, harmattan, sirocco, solano, Santa Ana—native to the places where they blow.

Some winds change the face of the Earth, working over the eons, subtly hurling grains of sand against rock faces, sculpting out magical shapes—pinnacles, needles, and arches. Winds move huge sand dunes, tons of mass, grain by grain, in sinuous ranks. The magnitude of the wind as an Earth-shaping force is plain in phenomena like windblown soils in one hemisphere that were first ground to powder by glaciers in another.

In addition to flecks of inorganic matter, winds carry other small particles—pollen and spores—moving germs of life around, colonizing new places, and abetting the great cause of genetic variation. In the back eighty, I can watch this transport as autumn gusts gather up seedheads of switchgrass and tumble them in the air. The wind delicately lifts them high overhead, then piles them in blond thatchy drifts against wire fences and borders of close-set firs. It carries wisps of milkweed floss, causes leaves to tremble, rips them free, whirls them up in little cyclones, and, after a short dance, abandons them to the Earth to settle and begin more soil.

Bob Dylan, the great bard of the sixties generation, sang that you don't need a weatherman to know which way the wind blows. True enough. If wind there be, you can just step right out and feel it on your face. Your breeze is a local phenomenon, local as the canyon between the high-rises or the little hill you're perched upon, local as your cheeks and their nerve endings, as the fine hairs clothing your skin. The wind that spawns the breeze may come from half a world away, however. We live in an era where technology lets us see, from twenty-two thousand miles up, which way the wind blows across the hemisphere and around the planet.

Although satellite imagery has provided us with the big picture of weather systems, it hasn't entirely supplanted land-based observers who check rain gauges and anemometers and phone the results in to a network. Other people whose lives depend, one way or another, on the dynamics of the weather—farmers and sailors—are also keen readers of the signs written in the shapes of the clouds, the hue of the dawn, the shifting of the winds. And many of us, with no training at all, can anticipate some barometric pressure changes simply by feeling them in our bones.

The Gaia Hypothesis asserts that Earth's atmosphere is continuously interacting with geology (the lithosphere), Earth's cycling waters (the hydrosphere), and everything that lives (the biosphere). Evidently the atmosphere has always been an integral factor of the evolution of life. It seems not to have been passive, but an active chemical medium participating in evolution. The atmosphere has been changed dramatically by the appearance of certain life-forms (the cyanobacteria, for instance), which, incidentally, created conditions more favorable to other, subsequent life-forms—like us.

Thus the hope is that the living air itself will offer some forgiveness: Gaia's autopoiesis. (Autopoiesis means self-making—"The concept is that it is intrinsic in cells and organisms to maintain their organization via interactions with their environment," explains Gaia Hypothesis coauthor Lynn Margulis.) The image is that the atmosphere is a circulatory system for life's biochemical interplay. If the atmosphere is part of a larger whole that has some of the qualities of an organism, one of those qualities we now must pray for is resilience.

FIRE

For our planet to be just the right distance from the sun, to be reached by its light in just the right measure, to be warmed into life and neither baked nor frozen into an eerie abiological silence or chaos, is a remarkable happenstance. A perfect sunny location is the happy fate of the Earth, our world, so unlike any other that we have seen. Yet there is fire at the heart of the Earth, just as fire is the heart of our solar system, and the heart and hearth of human culture.

Fire animates and transforms, frightens and fascinates. The human relationship with fire is unique to our species. Humans found fire long before they made fire—perhaps a lightning-struck wildfire, a nervous red line trailing billows of smoke, chewing its way across the savanna. It's definitive that humans are fire-using animals, tool-using animals. Fire can cook meat, bake bread, glaze a pot, forge steel, and keep the beasts at bay. The campfire is an intimate sun in the night, dispelling the dark, consuming the fuelwood, leaving little in the morning but black coals and gray ash.

Even in people to whom the sight of a fire in a circle of stones or in a hearth is rare, there is some deep knowing that this mastery is the essence of our power as a species. Fire has been one of our most ancient tools, a means of keeping openings for game, of clearing little plots for slash-and-burn agriculture, and of creating prairies where favored wild plants could dependably be harvested. By and by came fire broiling a planked salmon under wet lowering skies. Next an ember from an old home coming to sanctify a new hearth. Then woodsmen going for the fire locked in ancient oaks. And now fire unleashed from anthracite coal, from crude oil, from fuels laid down millions of years ago, those famous fern forests abuzz with their three-foot dragonflies. Fossil fuels yielding fire so hot it can melt steel.

Momentous changes came with our capturing of fire. Invention became a prime activity steadily gathering momentum, but it wasn't until the development of the steam engine, around 1769, that fire could be translated into motive power. Prior to that time, wind and falling water had powered mills, but those energy sources, although renewable, weren't portable, or necessarily constant. Thus the steam engine was a technological breakthrough of inordinate magnitude. About sixty years later the invention of the locomotive followed and inaugurated our modern era of transportation, centralization, and haste. Our burning of nonrenewable fuels has accelerated wildly over the last two hundred years, especially during the last fifty years, not merely changing the face of the Earth (and, with it, human culture) but also the very composition of the atmosphere.

We've come a long distance from those tentative, chancy beginnings of our relationship with this element. Fire still threatens to consume us—or rather by our use of it we threaten to consume everything. Today the fire in the chambers of our internal combustion engines is perhaps the single greatest culprit, although the technologies that preceded the automobile literally paved the way. Perhaps the worst thing we've done with fire has been the greenhouse effect. Primarily from our burning of fossil fuels, we are adding carbon dioxide (and several other "greenhouse" gases) to the atmosphere. (For instance, an automobile driven ten thousand miles a year will produce its own weight in CO_2 annually.) This increased concentration of greenhouse gases is trapping more solar warmth. The consequences of this global warming will beggar the imagination, and only radical change in our patterns of energy use can mitigate them.

Behavior counts. Usually behavior counts way more than belief. You can believe that the environment is in crisis and that it's important and that something should be done, but if none of the doing is taking place in your personal sphere, then nothing is changing. Because the planet is, in certain senses, finite, and somewhat of a closed system, we all live intimately with the results of our physical acts. Things do add up, and as population grows, there are more of us adding to the adding of things. Similarly, the benefits from many individual actions of self-restraint, frugality, and material simplicity will add up. It won't be easy, but it will be necessary.

Extraction of energy resources has always exacted a heavy toll from the Earth, beginning with the deforestation of the Mediterranean basin, not just to build ships, but to melt the metals that named the Bronze Age. Coal mining was another killer, first drawing men deep into the Earth, into dangerous toil, and then in our time, with the advent of strip mining, scraping bare whole counties and sacred sites to get at fossil sunlight. Uranium extraction for the nuclear economy displaces traditional peoples, exposes miners to deadly radiation, and inaugurates a wildly expensive energy-generating system whose byproducts are toxic for millennia.

If global warming weren't enough to persuade us to phase out the burning of fossil fuels, there remains the fact that on a finite Earth, nothing exists in unlimited supply. Sooner or later we will run out. With the case of Alaskan oil, in light of the dreadful environmental consequences of removing and transporting it from that fragile environment, why couldn't we just act as though it were gone already?

The sun beams ample energy to planet Earth, in a form that is perfectly useful to a lot of living organisms—the plants. Solar light makes grass and flesh. It "causes" photosynthesis, that primal source of all fuels but nuclear. Sunlight can heat a house in winter and charge a battery to run a fan in the summer. Environmentalists have long been fond of saying that the sun is the only safe nuclear reactor, situated as it is some ninety-three million miles away. The Earth's orbiting the sun, the Earth's tilt of axis, vary the length of our days, change the seasons, drive the climate—wind and ocean currents. Earth's dynamic all proceeds from the sun. Our ancestors understood that and were far more sensitive than we to what they perceived as the sun's comings and goings, observing a yearly cycle of holidays at the solstices, equinoxes, and cross-quarter days. Nowadays alertness to the sun's track through the sky can be indispensable to someone siting a home to take greatest advantage of solar light and warming, or building a greenhouse, or locating a garden.

Architecture and urban design can once again follow the track of the sun and defer to the weather to reduce the need for energy to provide heat and light, and to restore the strength and joyousness of human mobility, by foot or pedal power. Also, moving toward simpler, less wasteful household economies with all due speed will help greatly.

For urban dwellers, that planning and building will need to take place in the social sphere first—neighborhood organizing, car pools, cooperatives of various kinds, and working with municipal governments to subsidize conservation retrofits for low-income families, for instance. We could also respectfully suggest that there need not be a blazing light for every broken heart on Broadway or in Las Vegas and agitate for a less gaudy, more elegant, and energy-conscious aesthetic in our cities.

Frugality might become the latest style, simplifying life by gaining mastery over one's wants. Such action, combined with lively civic participation, aimed at "greening" the city, is a worthy task for the creative intelligence, and very human—like gaining mastery over fire.

WATER

"It's true that they have 365 faces a year," remarks my friend Debra as we drive by a lovely little lake in early October. Dawn is just beginning to color the sky, and is mirrored in the water's still surface. Tendrils of mist rise from the lake, whose temperature is warmer than that of the air on this chilly morning. A little island clothed in pine trees stands in dark silhouette against the water's luminosity, which joins that of the horizon. Just the sight of water can be soul soothing. An hour later, stroking through the bleachy blue waters of a municipal pool, I'm thinking, water is so smooth, marveling at how almost not-there it feels against the skin, how kindly it buoys one up, how sparkling and yielding it is. About two-thirds of our weight is water. It's the basic constituent of all our tissues, of the fluids that course through our circulatory and lymphatic systems, of tears running down our cheeks, of saliva, mucus, urine, and amniotic fluid, all the waters of bodily life.

Water is so everywhere-present, and such a commonplace, that its gifts seem almost limitless. Water is the only substance on Earth that exists in all three physical states—solid, liquid, and gaseous. Make it hot enough and water turns to steam, a gas with enormous expansive force; make it cold enough and it crystallizes, becoming snow or ice—water you can walk on. It moves gracefully from state to state. Water's physical properties are extraordinary, and its relation to Earth is integral—the biosphere runs on water, sunlight, and minerals. All these ingredients are present on the other planets of our solar system, but the temperatures of those planets are too extreme, too hot or too cold, to allow liquid water in its solvent, circulatory form. Flowing water is a presence unique to Earth.

On Earth there is so much of it in fact that 71 percent of the planet's surface is ocean. Because of water's extraordinary capacity to absorb and release heat slowly, sizable bodies of water, primarily the oceans, temper climate, maintaining a range of temperatures congenial to life. This tempering is a gift of elemental grace. Ninety-seven percent of the Earth's water is in the salt seas. Most of the fresh water is locked up in the polar ice caps. Thus, less than a percent of the water on Earth is available as fresh water.

Gravity eases it down, solar heat drives it up: precipitation, evaporation, and evapotranspiration continually cycle water through the atmosphere and biosphere. From the deep rolling immensity of the oceans, from the surfaces of lakes and streams, the air takes up water; clouds are formed; dreamy, delicate scarves, massive thunderheads ominous with precipitation—numberless, shifting forms. Some rain clouds slake the Earth gently; others lash with downpours. Fog may simply accumulate in a forest's foliage—like the needles of a redwood tree—and drip quietly to the duff below.

That water always flows downhill is axiomatic; the rate at which it does is determined by the condition of the land on which precipitation falls. Vegetation buffers the impact of rainfall, increases the porosity of the soil and its absorption of water. Roots and other organic matter invite water downward. Rain or snowmelt may percolate through the soil into the Earth's crust. Precipitation landing fast and furiously—in volume surpassing the surface's capacity to soak it up— qualifies as runoff. When the land's ability to slow runoff is diminished by being paved, logged, strip-mined, overgrazed, or too heavily farmed, erosion and flooding result.

Land use upstream conditions water's way back to the sea, and the fate of all the ecosystems en route. For instance, careless logging upslope speeds runoff, which scours out and silts up the gravel beds where salmon spawn, destroying fisheries. Paving paradise and putting up a parking lot (a lot of lots, actually) seals the surface, giving water nowhere to go but to overload sewer systems after every big rain.

In undamaged watersheds, precipitation cycles back to the sky by two pathways: water soaking into the soil, dissolving minerals, is taken up by the roots of plants, moves up their stems by capillary attraction, and out through pores in leaf surfaces to evaporate into the stuff of clouds—evapotranspiration, it's called. It makes for an intimate mutuality between forests and rain. Precipitation falling on steep highlands trickles downhill to join rivulets which become creeks which feed streams which join rivers which flow madly or lazily to the sea. And from the surface of the sea, at the touch of the sun, clouds of water vapor will rise to begin the cycle again, water moving everywhere in the biosphere, the definitive element of earthly life.

Because of our absolute need for water, human settlements are often found near rivers, on floodplains, and at harbors. As a result of land degradation upstream, floodplains are increasingly chancy places to dwell. The floods that inevitably will inundate them naturally instill a primordial fear. Perhaps that is why the myth of a world flood is a part of many cultures' telling of their distant past. Water has its ominous qualities too. Flash floods or drowning are the stuff of nightmares—first the water reaches your windowsills, and then your roof line. What becomes of you, clinging to the roof? Not a deluge, but a gradual insidious flooding may be the result of global warming, which is melting ice at the poles and causing a rise in sea levels. With a rise of only a few inches, coastal settlements may be encroached upon and low-lying islands entirely submerged, catastrophically leading to the displacement of millions of humans. Rising sea levels also could cause the salination of low-lying water supplies.

Water in the shape of rivers gives us images of endless change and flow, ever same and ever changing; some rivers placid, some raging, some broad and muddy, some rocky and hissing, pathways back to the headwaters for salmon and shad, sturgeon and smelt.

One of the truths most clearly expounded by watercourses is that geometric human political boundaries have an arbitrary quality, cleaving right through watersheds, lake basins, and river systems. These territorial divisions act to the detriment of the waters, reinforcing the misconception of a separate, autonomous existence in persons and peoples; the delusion that it is a state's or nation's prerogative to deforest its uplands; or to dispose of contaminants in, or to overdraw on, its river systems, regardless of the effects on all the lives downstream, and throughout the body of water where they are bound, be it an ocean or an inland sea.

If the ecology movement has a global totem, it is probably the whales. The seas are zones of natural mystery. The enormity of whales, their dwelling in the deeps, their songs, their society, their grace, and their possible great intelligence haunt us with the thought that we may not be alone on this planet in having consciousness. A consciousness shaped over millions more years than ours has been, and in a different element, may, although ultimately unknowable, merit some deference and should command some protection. Sadly, after decades of struggle to put an end to whaling and to preserve the few percent of great whales remaining, outlaw nations still refuse to agree to a moratorium on whale killing.

The plight of the whale epitomizes the modern dilemma, the rift of mind that underlies the ecological crisis. To some humans, whales are just a resource, a form of protein, like cattle. They can be eaten, so why should they not? Why shouldn't the investment in factory ships be amortized, even if the "resource" is annihilated in the process? People have to work, and people have to eat. Meat is meat. Although leviathans have captured the imagination of school children and adults all over the planet, the apparent failure of this global outpouring of concern to secure the whale's evolutionary destiny is shameful. What do we have to do to make it come out differently?

Careful stewardship implies a fundamental respect. The tradition of regarding the Earth as a mother suggests the correct attitude: not presuming on the Earth's great generosity, but to become quite sparing in our use of her blessings, and grateful for the very fact of our being, which, like flowing water in the solar system, is a rare gift.

SPIRIT

It goes by dozens of names, and different people see it expressed in all life: something larger and more mysterious than the individual self, often called spirit. Our human conception of soul and psyche is largely shaped, not to say limited, by the human mind. Now that the image of Gaia has become part of global culture, however, and our understanding of ecology has deepened with this process, it seems the life force itself shows intelligence: mind in nature.

Everything connects. We may not all be aware of it, but we do interact with all life; and we all bear some measure of responsibility for the future of all life. In this age of scientific information and spiritual exploration, the necessary knowledge to effect change on behalf of the life that lives through us and all beings, the life through which we live, has never been readier to hand. Whether there is wisdom enough remains to be seen.

And at this turning point in life's history on Earth, the human variety of consciousness is both a blessing and a curse. It was mind (and opposable thumb), after all, that got us into the current mess, and mind's gifts to life—human creativity, community, cultural diversity, learning, love, spiritual aspiration, and art, to name but a few—could lead us out.

A species-wide soul crisis has precipitated a global ecological crisis, and this offers us the opportunity to mature. A sadder but wiser species, what we are here for now is reclaiming, restoring, preserving, protecting, atoning. Now we have a chance to become more real to ourselves and to see ourselves in full dimension.

Human striving for increased consciousness and compassion is a limitless frontier: it embodies faith, dogged determination, altruistic actions, believing in your beloveds when they've lost belief in themselves, striving to experience oneness with the universe, studying to divine the meaning of sacred texts and sacred syllables, fearless surrender to the infinite, epiphanies on the shore. This yearning, latent in everyone, is indomitable despite the privation and alienation into which the human species is sadly plunged.

When my work on this writing was just beginning, I'd been browsing through a color atlas of Hopi Kachina "dolls." (Like a lot of folks whose attitudes were ordered [or disordered] in the sixties, I am captivated by the story of the Hopi people, an American Indian tribe that for centuries lived a sacred, sustainable lifeway in northeastern Arizona, a remote, starkly sublime, arid, and exacting region.) The mesa-top Hopi pueblos are the oldest continually inhabited settlements on the North American continent. Notwithstanding all the past and present threats to their cultural survival, traditional Hopi emissaries have traveled the world, especially in recent decades, to share their spiritual understanding of the nature of balance and to speak for their right to continue as a people. The possible loss of this or any culture's earthly wisdom is grievous to all cultures. The global community needs not only new knowledge to get there from here, but attention to the old ways, too.

In Hopi understanding, Kachinas are not exactly gods, but indwelling tutelary powers in a myriad of plants and animals and phenomena that are the world of the Hopi. "The basic concept of the [Kachina] cult," wrote Barton Wright in *Hopi Kachinas,* "is that all things in the world have two forms, the visible object and a spirit counterpart, a dualism that balances mass and energy. Kachinas are the spirit essence of everything in the real world...." Listen to the English versions of the names of just a few of the Kachinas: Antelope Kachina, Assassin Fly Girl, Badger Kachina, Big Ears Kachina, Blue Corn Girl, Broad-faced Kachina, Butterfly Man, Chasing Star Kachina, Chili Kachina, Cloud Guard Kachina, Cocklebur Kachina,

Crow Mother, Cumulus Cloud Kachina Girl, Dung Carrier, Great Horned Owl, Hand Kachina, Hump-backed Flute Player, Lightning Long-haired Kachina, Meteor Kachina, Prickly Pear Kachina, Star Kachina, Sweet Cornmeal Tasting Mudhead, Water Serpent Kachina.

"The Hopi do not worship these Kachinas," says Wright, "but rather treat them as friends or partners who are interested in Hopi welfare. Because it is not easy to interact with the Kachinas in their insubstantial form, it remains a simple matter to give them shape and personality by impersonating them." During the year the Kachinas visit the Hopi and make their presence felt in a ceremonial cycle of dances and songs that are performed by members of sacred societies and attended by the entire community.

"The men who impersonate Kachinas and dance in the plazas carve small wooden replicas of their Kachina appearance and present these to infants and all ages of females," Wright continues. "This carved and painted figure is called a *tihu* by the Hopi and a Kachina "doll" by others. It is not a doll, a plaything for children, but an effigy or small part of the Kachina it represents...." The imagery of the costumes and *tihus* is not literal but strikingly abstracted, a visual intuition of indefinable things that is heightened in significance by the symbolic content of every detail, from the colors in which they are painted to the different kinds of feathers that once would have adorned them.

So the Kachina *tihus* are tokens of the powers that determine the world, and the Kachina ceremonials are occasions of teaching as well as reverencing. Everybody in the pueblo instructs the young, and the message is that one does not live merely for oneself: one lives for the people and the continuity of their way of life. Living in a rigorous environment (which, through human agency, all environments are becoming) demands discipline. And if this discipline is imposed latterly or externally, it's likely to be an ill fit. So the genius of such a traditional people is that its lifeways are taught continually. All elders in the tribe are expected to convey the right way of life by a variety of means, from cradle songs to epic creation tales to ritual offerings in sacred places to the use of symbols to encode parables in familiar places and common objects; the way fit to the place. In doing this, the Hopi culture is making sound use of the human gifts of memory, language, artistic creativity, and humor. Within this integument of culture, the tribe is itself a being inseparable from its surround.

However much I may long for something like the sacred knowledge of a tribal people like the traditional Hopi, though, I can't somehow become one. Rather, I have to reflect and act on what movement toward spirit-felt Earth-reverence can be attained in my own culture. Ruminating on this one mid-October morning, I went into town to watch a friend teach a children's aikido class. This far north, it was just getting light, and the sky was full of dingy, cottony rain clouds: the landscape heading toward winter. Much of the glowing gold and flame of the fall foliage had been dashed to the ground by recent rains and a snowfall. The new wetness was bringing out the intensity of the remaining colors and blackening the bark of the trees, defining their bare-branched identity against the neutral sky now that their cloaks of leaves were ripped away. Cresting over the hill, I saw the sunrise causing the underside of the clouds to blush briefly, a glory over the saturated teal of the bay's waters. There was a momentary epiphany in this, a glimpse of the magic immanent in an everyday sight of farmland, sky, and shore.

Onward I drove into the ordinary reality of the town on a Saturday, among my fellow citizens who were also out on errands or looking for a little recreation, and out to the gym at the community college. The aikido class consisted of maybe a dozen children, most wearing the aikido *dogi,* a simple, loose-fitting, white costume, donned and belted in a manner that connotes the principle of aikido: nonresistance, a way of finding harmony. There were four adults: my friend teaching; three others helping. The kids ranged in age from six to ten, and there were more boys than girls. It was touching to watch these young ones trying something new, something foreign. Here were modern kids tasting a discipline, taking the big visible risk of learning through trial and error. They all looked lithe and lean, compact and muscular as puppies, and were pretty much paying attention. Some were learning by observing; others were not quite getting it. It was a world, that class, aquarium-like under the harsh gymnasium lights. Looking at those young ones, I ached, knowing that, much as I or any of us might desire it, I can't fix the world enough to forestall the suffering their lives will inevitably entail. Life in her profligacy exacts a commensurate toll.

In *The Edge of the Sea,* Rachel Carson describes different oceanic and estuarian life-forms, creatures whose spawn may number in the millions. Among those millions, perhaps only a few will survive into maturity. Most of them will be eaten. The implacable destiny of the individual organism is first to live and then to die. There's a tragic edge to this reality. That those aikido kids, indeed all kids, are emerging into a world that will increasingly confound their parents and, henceforth, will challenge them to the utmost is pretty stark stuff. That none of us gets out of this one alive is starker still.

Quite apart from one's initial subjective and self-interested responses, gloom-'n'-doom prophecies raise a tough question. What can we bring in the way of spirit and psyche to times like these? Once, at a gathering that consisted of predominantly middle-class white intelligentsia, Chief Oren Lyons, Faithkeeper of the Turtle Clan, Onondaga Nation, Haudenosaunee, described concisely his people's lifeway and, out of that, offered his own counsel for coming times. The wisdom that struck me was that "much will be demanded of those to whom much has been given." Another thing he said was, "Don't look for mercy where there is none."

Ours is a pivotal generation. We have to do more now than just live our little lives. We've got to become strong and enduring and respectful and speak to those qualities in one another. Civilization's mishaps have brought us to a time for growing up. Not to despair and not to hate and not to frighten, but to engage one another compassionately, to confront lovingly, to embrace the entire reality, and not to gainsay the pain are some pieces of the task.

It might be said that spirit is the sublimest expression or the ultimate flowering of the survival instinct. The movement songs of the sixties all say you can't kill the spirit. Spirit, if it is, is invincible. But people are fragile, flesh and blood, discourageable. External circumstances work differently on different people, and there's no predicting how. Where one human may rise in creative outrage against an injustice or senseless act of destruction, another may sink under the pounding waves of want and frustration.

Despair, understandable enough in times like these, always threatens the continuance of body, mind and spirit. Simple ignorance and obliviousness, not knowing even a few details of life's infinite, marvelous patterns here in the biosphere and hence not knowing what there is to care for, is another deadly threat. Injustice, racism, alienation, indifference, violence, sexism, loneliness, greed; the destruction of traditional cultures, villages, neighborhoods, and countrysides; the rising levels of suffering that result from ecological deterioration: the bill of dispiriting particulars is too long.

Simply to deplore humanity's crimes against the planet and against itself is not a rhetoric in which to get stuck. Is it not essential to any other change, though, for us to acknowledge that we have perpetrated ecological atrocities, and to conjure with that culpability as a part of our common, and individual human heritage?

Then we can go forward, out from our experience as the creatures that simplify the biosphere, and begin to live in ways that are sustainable: materially spare, but spiritually rich. And we must take with us humility learned through the failures of our unending, often disastrous efforts to bend nature to our purposes. Members of industrialized societies could easily succumb to the psychopathology of control when confronted by task of reinhabiting a damaged planet.

It's infantile to escape into technofantasies of colonizing Mars with life or look to biotechnology to invent "life" forms adapted to our worst pollutions or hope for some sudden, collective, mass-mediated awakening (which seldom seems to specify any requirement to rein in our wants) to precipitate an abundant millennium. We must admit that we cannot run the natural world according to the dictates of the ego. This delusion has already cost the planet a half million of our companion species and thousands of traditional lifeways. A certain bounded experience of remorse, the inseparable shadow of prodigious hungers, inventions, and enthusiasms, might be seemly. The dominant minority on the planet could use some depth of soul.

Penitence might engender in us a passion to heal. Studying the ways of ecosystems so as to get the basis for a faithful regeneration of the land could reveal a holier writ. It would be a version of inviting the geniuses of place to tell us how they generate sustenance, raw harmony, and myriad embracing forms.

The Earth and its ecology are no mere notion. The world is real, and human ecological sustainability consists in hundreds of individual physical practices and disciplines, every one of which can be performed as a ritual of respect for the planet that gives us life. Going on litter patrol in the park, starting a compost bucket and maybe even a community garden to spade it into, learning the name and ethos of that little brown bird in the backyard, traveling on foot through a forest, patching the elbows on a sweater and making do with it a while longer—these can be preliminary disciplines and pleasures to commence more frugal lifeways grounded in respect for the Earth.

These soul-polishing labors and meditations are means of healing the human spirit. Imagine the sweetness and freedom in this reunion, the goodness in having a daily life that's integral to the realities of place and a worthy basis for culture.

136

Learning to provide for ourselves, to care for the places in which we live, and to restore them to biological health; learning to thrive using the renewable energy flows of the elements; studying the natural histories of our home places and discerning in them the outlines of our future self-reliance; consecrating large land areas across every continent as wilderness shrines where entire ecosystems may continue in their evolutionary destiny—all of these actions aim at becoming native again to place; bioregionalists call it reinhabitation. Reinhabitory humans make love with their home places, bringing forth a wealth of cultures, songs, images, teachings, inventions, musics, and cuisines. Part of the hope in spirit, paradoxically, is coming back to our senses, and being able to once again to revel in them.

Remember this: to be alive can feel like such a piece of good fortune as to leave one breathless. To have a pounding heart and nerves that crackle with information and a skeleton strung with muscles strong enough to pump the pedals on the bike or swing the maul to split a little wood; to have eyes to scan the valley and register the subtleness of the color change as September pushes fall forward is the ultimate gift, a consummate luxury. To experience this bliss in life can and should be every human's birthright.

In the invention and regeneration displayed in the wilderness and the renewal displayed in the turning of the seasons, Earth spells her hope for us. In the persistent human yearning after the good, the true, and the beautiful, and the mounting struggles around the world for a more decent, life-affirming society, are spelled our hopes for ourselves, hopes that are awakened today even in the very young.

One early afternoon, not so very long ago, a cadre of four little girls wound their way up my long driveway to pay a call. Having learned that the home planet is endangered, they had taken it upon themselves to visit all the neighbors in a home-grown canvass to save the Earth. They had made a creative, positive, and quite touching response to the prospect of planet doom. They were distributing hand-drawn inspirational posters. I got one that shows our blue and green planet rising in a field of eight yellow stars. "Save the Earth" reads the headline. "Pitch In. Help Out. If you're interested, call. Don't stop trying."

SOME WORDS FOR THE WILD

(2001)

Thirty years ago, when I was a youthful hothead, we decried the callous absurdity of a society that was knee-deep in garbage, firing rockets at the moon.

Today we're in the midst of the sixth great extinction crisis. The climate is changing even faster than scientists first warned. 1.1 billion human beings lack access to potable water, and the United States, or rather the ruling cabal of the United States, insists on building a missile defense system to protect us from some yet-to-be-named enemy.

About that extinction crisis: technology and globalization have, over the last half-millennium, done for the planet's biodiversity what the asteroid that hit the earth and put paid to the dinosaurs did. Given the imminent consequences of global warming for nature and humanity, it seems fair at this point to call anything that promotes increased consumption of fossil fuels heinous. Transoceanic and transcontinental trade, trucking, and mass travel by conventional means—the delivery systems of globalization—all come under this heading. The synergy of technology and globalization is ending the evolution of whole lineages of life and even threatens whole biomes.

The engines of destruction have been gaining momentum since 1692, when, legend has it, a Dutch sailor shipwrecked on an islet off Mauritius saw and, perhaps, ate the last dodo bird. It was an inadvertent thing, just a side effect of the navigation of the Indian Ocean in a quest for resources, markets, and territory. The sailing vessel was the technology of globalization that happened to end the dodo's evolutionary journey and enrich the idioms for deadness.

Today, according to the International Union for the Conservation of Nature's Red List of Threatened Species, one in four mammal species and one in eight bird species face a high risk of extinction. In the last 500 years human activity has forced 816 species to extinction or extinction in the wild. The rate of extinctions now is fifty times what it would be without benefit of modern economics and technology. "Fifty times" is an aggregated number, but the reality consists of distinct lives and traits and talents—the wonders of nature and the basis of life.

Today, like Thoreau, I wish to speak a word for nature, because nature's flourishing—which means our own—is under devastating assault. So I'm going to talk about some of those lives and traits and talents, to flesh in some of the details of what's going on out there beyond the human community.

Why do you suppose that Thoreau, that great American pencil-maker and sage, tax resister and bean planter, declared that "in wildness is the preservation of the world"?

It's because evolution is wild. This 4.5 billion year old planet has been, for all but the tiniest fraction of its history, wild. Of its own self-will, Earth has brought forth marvels: bowerbirds, coral reefs, compass plants, jellyfish, white pines, snow fleas, lammergaier vultures, root fungi, monarch butterfly migrations, 290,000 species of beetles, rainforests, prairies, sperm whales, wallabies, strangler figs, sturgeon, purple lady slippers and *Homo sapiens*. Natural barriers like oceans and mountain ranges promoted this wild variety of life forms, this natural diversity and cultural diversity and its fruit. The adaptations to place and circumstance and the vast web of relationships amongst all these creatures are astonishing, but mostly disregarded.

Consider the yucca and the yucca moth. This is a story of seed for seeds. Most insects pollinate flowers incidentally, on their way to the nectar. In some species of Chihuahuan yucca and yucca moths, though, the female moths have specialized mouthparts with which they collect pollen from the yucca flowers. They gather pollen and stuff a big ball of it into the yucca flower's stigma, the part that leads to its ovary, which holds the future yucca seeds and is also the place where the moth deposits her eggs. Thus a portion of the fertile seeds will feed the larvae, the future yucca moths. Insect pollination as deliberate as this is quite unusual. The moths and the yucca depend on each other. When the moths are absent, the yuccas don't set seed and soon die out. And when the yuccas are absent there are no yucca moth nurseries. Such vital partnerships are everywhere in the wild and generally overlooked when CEOs and their henchpeople in lab coats try to streamline nature for profit.

Although Thoreau undoubtedly never saw a Chihuahuan yucca, he was a keen observer of the forestry practices of his neighbors the squirrels and jays, and understood their collaboration with oaks and pines. All the free self-willed phenomena he saw during his sauntering around Concord must indeed have declared to Thoreau that in wildness is the preservation of the world. Wildness tells us that everything is hitched to everything else. But the extinction crisis is rapidly uncoupling myriad vital relationships and fraying the fabric of life on earth.

We may take from Thoreau that we should hearken to our own deepest experience. As a species, human beings have more experience being wild, living as semi-nomadic subsistence peoples embedded in healthy ecosystems, than in any other lifeway. For about ninety thousand years, wild was us. Agriculture has been going on a mere ten thousand years; reliance on fossil fuels a hundred and fifty years, and only in the last thirty years has human life has come under the influence of computer-based technology. Subsistence is given by the whole web of life. But in the accelerating drive to subjugate nature to profit—the narrowest of human purposes—technologies that are blundering in the extreme and global in their effects are unleashed.

In the manufacture of paper, plastics, pesticides, and refrigerants, the chemical industry has managed to pervade the tissues of most higher animals on the planet with persistent organic pollutants (POPs). Many of these chemicals mimic hormones. They interfere with the reproduction and development processes of many vertebrates. Such hormone disrupters have induced subtle damaging changes in the behavior and intelligence of children born to mothers whose diet included fish with such POPs concentrated in their flesh. In some places, alligators, cormorants, and gulls have hatched with deformed gonads. Are learning-disabled kids and impotent wildlife a reasonable price to pay for profit, or even progress?

Nobody knew or thought to ask whether propellants like freon would eat ozone. But they do, and extra ultraviolet radiation exposure is a consequence. Increased UV exposure seems to be among the reasons that frog populations around the world are crashing. Nobody even knows where trifluoromethyl sulfur pentafluoride, a greenhouse gas 18,000 times more potent than CO_2, comes from, but it's in the atmosphere doing its warming work now.

Meanwhile, the trade geniuses are arguing against the precautionary principle. That is the sensible idea that new chemistries, technologies, and other tamperings with evolution should be presumed hazardous until proven innocent. Economic activities like long-wall coal mining, gold extraction by cyanide leaching, clearcut forestry, petroleum exploration, drilling, and transport do their visible, immediate damage locally, but in sum they also accelerate a planetary extinction crisis. Unbridled profiteering at the expense of the wild has been effecting ecosystem-scale damage for centuries. The world trade in furs did a job on the North American landscape by decimating its beavers, of which there were an estimated 60 million in 1492. With their appropriate-scale, low-tech dam-building talents, beavers renewed the landscape. Their pond-making, water filtration, and meadow creation led to a rich patchwork of habitats. Tough luck for the beavers that in the 18th century beaver hats were all the rage. In just one year, 1743, just two shipments of furs to France included over 150,000 beaver skins between them.

The modern era, then, with its relatively crude technologies like rifles, steel leg-hold traps, double-bitted axes, two-man saws, ox-drawn sleds, canoes, railroads and steamers, spurred the extinction crisis. Wildlife and wild habitats in the "New

World" were already depleted and fragmented going into our era of megatechnology and globalization. Today's technologies of mass commercial transportation—jets, cargo ships, tankers, superhighways, and trucking, even electrical transmission lines and towers—are intrinsically destructive of ecosystems. For instance, a conservative estimate finds that each year 80 million birds die due to collision when power lines and towers impede their natural migratory cues. Larger birds are simply electrocuted. Several of the newly released California condors, fetched back from the jaws of extinction by a captive breeding program, met their doom in exactly this way.

No wonder there are Luddites still among us!

———

In a famous article in the April 2000 issue of *Wired* magazine, Bill Joy, a cybergenius who describes himself as "an architect of complex systems," published "Why the Future Doesn't Need Us," which created a serious buzz. In it, Joy worried aloud about the imminent advances in, and synergy of, robotics, genetically engineered organisms, and nanotechnology. These technologies share, he said, "a dangerous amplifying factor: they can self-replicate." Thus we may have to confront a technology that is not just *autonomous*, to use Langdon Winner's formulation, but *autopoetic*. Joy feared that certain applications of these technologies might be capable of reducing life on the planet to "gray goo," if, say, a rogue nanobot with a mind of its own and a bad attitude flew the coop.

We have good historical grounds for expecting bad things to result from the introduction of alien organisms to innocent environments. In addition to being the vehicles for extracting resources from the New World, the sailing vessels that launched the modern colonial era were vectors for a biological invasion of the new world.

Local extinctions are a common result of biological invasion. And "local" can be as large as Lake Michigan. Here's a for-instance. The Great Lakes basin's aquatic fauna was revolutionized by commerce and transportation in little more than a century. First over-fishing stressed the system, and then the construction of the Erie and third Welland canals allowed the sea lamprey to enter the Lakes and dine heavily on the native lake trout, which were top predators. Too bad, because the lake trout could have come in handy to eat up some of the millions of invasive alewives. The alewife is an oceanic fish that entered and naturalized in the lakes, but with limited success. Occasionally at spawning time, the alewives, poorly adapted for life in fresh water, will die by the millions, leaving Lake Michigan's shores piled with drifts of rotting fish. Meanwhile, the sea lamprey were beat back with poisons and spawning barriers, and Pacific salmon were introduced to fill the niche of the lake trout. The salmon aren't reproducing naturally in the lakes, but the sport fishers like to catch them, so Michigan continues to stock the lakes with hatchery fish. Hatcheries, which are basically aquatic feedlots, are themselves problematic because they produce nutrient-laden effluents that degrade water quality. That tragicomedy of ecological errors gets us from the mid-nineteenth to the mid-twentieth century in the Great Lakes.

Late in the twentieth century, just 13 years ago, came zebra mussels, lurking in the ballast waters of some vessel that passed through the Caspian Sea. These little molluscs are so enormously prolific that they smother and out-compete native mussels and hamper their reproduction. Zebra mussels gum up the works of lakeside cities by encrusting and clogging water intake pipes. They are so numerous that their filter feeding is making the water too clear, which may be causing blooms of blue-green algae. These algae blooms make life difficult for still other old-time residents of the lakes.

It seems only fair to ask whether, with regard to technology and globalization, we know what we are doing. The sorry state of the Great Lakes ecosystem is just one piece of evidence arguing that the answer to that question is no and hell no. Bear in mind that all of those biological upsets were accomplished without benefit of NAFTA, GMOs and nanotechnology.

The commodities, conveniences, and consumer goods whose mass production is the justification for much of this technical innovation and trade liberalization are marketed now on an unprecedented scale. There are six billion of us. Whereas, just at the turn of the last millennium, there were a paltry 275 million human beings, and most of the world had yet to suffer the onslaught of what the West was pleased to call civilization. So the sheer magnitude of the human population accounts, in part, for the magnitude of the effects that the planet is experiencing.

Notice the two opposing patterns: globalization leads to simplification and instability, not to mention ugliness and misery; evolution leads to diversity, dynamic equilibrium, life, and beauty. Structural alternatives like relocalization can favor life's flourishing. Locales have always outlasted empires. And the collapse of empires is the rule, not the exception.

The vision and power and will to resist corporate imperialism and its technological tyrannies, and to implement diverse alternatives, desperately need to be kindled. Most large-scale human interventions in the planet's biology have worked like giving a Swiss watch a tuneup with a sledge hammer. People are beginning to understand this, but assume as given that you can't stop progress.

———

As he pondered the Promethean technologies he saw emerging, Joy looked to the reflections of Freeman Dyson, a theoretical physicist, on the development and use of the atom bomb, certainly one of the most horrific manifestations to date of what Dyson himself termed "technical arrogance."

The race to mutually assured destruction began with the bomb. Dyson said that "the reason that it was dropped was just that nobody had the courage or foresight to say no."

Nevertheless, at the end of his article, Bill Joy comes to the conclusion that "the only realistic alternative"—to gray goo, I guess—is to have the courage to say no. Joy proposes *relinquishment*: to limit development of the technologies that are too dangerous by limiting our pursuit of certain kinds of knowledge.

Is it too late to direct those inquiring minds back to the task of learning the plants? Or too late to promote the idea that it is not knowledge that's scarce, but wisdom and the wildness that engenders the flourishing of human, and more than human life?

Fun While
It Lasted

(2006)

I. Could This Be the End of Civilization
as We Know It?

My friends will tell you of my peculiar fascination with doomsday scenarios. To live in expectation of this civilization's collapse, which I do, can be isolating. It's like having historical-epochal halitosis. You don't always get invited back for dinner after (once too often) you've tried to engage the guests on our imminent decline and fall.

I didn't start out apocalyptic. As a baby boomer born into an affluent Sunbelt family, I assumed that the hedonic summer sung by my beloved Beach Boys would stretch on and on. In the late sixties, though, while I was in college in the San Francisco Bay Area, ecological awareness crashed over me like a series of rogue waves. There was a buzz about the 1968 bestseller in which Paul Ehrlich pointed out that the fuse on the population bomb was burning short. Also, about then, Paul Shepard smartly characterized ecology as "The Subversive Science," clearly implying that to acknowledge interdependence with all of life subverts any claims by *Homo sapiens* to be superior, or even exceptional, *vis-à-vis* other species. And then, in 1969, a Union Oil well blew out in the

Santa Barbara channel, washing the horrible pathos of oil-soaked grebes, cormorants, seals, and dolphins onto miles of Southern California beaches and thence to the front pages. So, about the time I got my diploma, I also got that my way of life was, in terms of land and other life, exorbitant. I began to question whether present-day civilization, which seemed to be running on a bum epistemology, was a good thing at all.

Over the decades since, ecologically concerned scientists and scholars have singly, or in learned conference, been issuing deadlines for humanity to change its ways. From 1970 (when the Club of Rome presented *The Limits to Growth*) to 2005 (when the United Nations's *Millennium Report* came out), these inventories of the state of the planet's bioregions and its remaining resources mostly tally losses and, for their pains, lose audience share. The publication of the *Millennium Report* didn't even make it into my hometown paper but, to be fair, the NBA playoffs were under way at the time.

Scratch a fear, find a wish. Although the archetype fulminates in many a subconscious and might, given nuclear weapons, become a self-fulfilling prophecy, the fall of a civilization is not necessarily the apocalypse. But unless you're adapted to the peasant or foraging lifestyle, simple collapse might feel apocalyptic enough.

Early on, civilization displaces oral tradition and subsistence lifeways, then grows by centralization, specialization, and exploitation of its hinterlands. By the time a collapse occurs, the ways and means of doing large-scale, sophisticated things like sustaining urban and suburban populations are lost. Subliminal awareness of that pattern may explain why so many devotees of Tom Brown's *Tracker* books and flint-knapping clubs are around these days. Quite a few people must want the ability to survive in the bush or to make real, rather than virtual, cutting edges, even when metallurgy has passed into the realm of the ancestors. But think: what would happen to all of us if Gore-Tex and Polarfleece fabrics were no more? Dressed any hides lately?

Collapsewise, the worst-case scenarios for Y2K provided a teachable moment, though if the worst had eventuated and nuclear and chemical plants all over the world had gone critical, the opportunity to say "I told you so," wouldn't have been a comfort to any Luddites strewn among the dead and dying.

I live in what Richard Nelson, the author and anthropologist, calls a "woodburb" outside Traverse City, Michigan, one of the fastest-growing areas in the state. Most of us live out here to avoid having to deal with our neighbors. In addition to doing a little writing and speaking on the subject during the run-up to what fortunately proved to be Apocalypse Not, I got a little cash out of the bank, put batteries in the flashlights, sanitized the bathtub for storing water, and made sure I had plenty of canned goods and firewood. In the interest of convivial, long-term persistence, organizing my neighborhood would have been the most meaningful thing to do. However, for this introvert it would have been the most difficult. In addition to my shyness about engaging in discussions of big risks, to forsake the company of like minds felt like the biggest risk of all.

Hard on the heels of the Y2K prospect comes now the factor that, interacting with other monstrous problems like climate chaos, actually could precipitate the collapse of global civilization. "Peak Oil"—the end of cheap oil—may slay the growth monster and, with it, the comforts most North

Americans take for granted. For us consumers, things will go hard as cheap energy becomes a sad memory. It'll be goodbye to the private automobile, the eighteen-wheeler transporting January strawberries, and the cornflakes; farewell to the air-conditioned combine in the county-sized cornfield; adios to natural gas-derived fertilizers and chlorinated hydrocarbon pesticides. Cancel the gym membership: it will be superfluous, post-Peak. Practice bundling up, freezing in the dark, taking the bus, and eating less. Instead of debating which electronic doodad to buy on credit at the mall this Saturday, we may be spending our underemployed hours roaming the countryside in search of work, food, fuel, or salvage.

About twenty years ago, I was in a head-on automobile collision. Phil, my husband at the time, and I were headed downstate to do some bioregional organizing. Phil was driving my compact Chevy Capri. Not far out of town, he pulled left into an uphill passing lane to get out from behind a slow semi truck, whereupon we discovered that another car was in that very same lane, headed downhill and straight for us. We were bound to crash. Things do go into slow motion at moments like that. I looked over at Phil. Our eyes met in a stunned farewell. The next thing I remember is looking over at Phil again and our eyes meeting again in the amazing discovery that we were still alive, even if in a hissing wreck by the side of the road. Although it took years for each of us to recover from our respective injuries, we did.

Almost twenty years later, early in 2002, I picked up my first copy of Richard Heinberg's brilliant *MuseLetter*, colliding with his series of articles on Peak Oil (later issued in book form by New Society Publishers as *The Party's Over* and *Powerdown*). Reading the *MuseLetter* on Peak Oil and its possible consequences was not unlike being in the passenger seat of that little Chevy compact, confronting an oncoming vehicle and discovering that the best I could hope for at the moment was to learn some physics and anatomy the hard way.

With much research and reasoned analysis, Heinberg details our society's inevitable collision with the limits to oil and natural gas production. They are close to zenith and will decline; he explains how utterly dependent much of the world's economy, particularly agriculture, is upon fossil fuel. I've always believed relocalization to be a desirable change, but living through a rapid default into it could be rocky. The serious demand-side responses exacted by Peak Oil: households and entire regions growing most of their own food, sticking to home, engaging in cottage industry, walking everywhere at a tenth of the speed of driving, inhabiting smaller, tighter houses in compact settlements, and practicing extreme frugality; while federal taxes are diverted from armaments to railroads and other public transportation—will take courage, work, and skills both bygone and new. These latter can be difficult to acquire in a crisis. A way of life is something more than an easy-to-master assemblage of techniques.

II. Peak Oil, Pipe Dreams, and Persistent Pessimism

"Peak Oil" is a catch phrase now in use beyond the circles of the petroleum mavens who coined it. In the mid-1950s, hoots, jeers, and denial greeted petroleum geologist M. King Hubbert's prediction that U.S. oil production in the lower forty-eight states would follow a bell curve and peak around 1970. Since then, data-gathering and modeling have become more sophisticated, as have oil prospecting and drilling technologies. Nevertheless, although some economists still contest it, worldwide oil production does look to be following Hubbert's curve to peak sometime in the 2000s. The easy pickings enjoyed by John D. Rockefeller and his colleagues are history. Oil exploration and extraction costs in money and energy will likely increase from here on out. And because demand for oil continues to rise—not just in America but among the burgeoning motoring middle classes in India and China—the remaining global supply of oil is apt to go quickly, if not overnight. Unless an international oil depletion protocol is adopted, oil will be a *casus belli* until the lights go out all over the world.

On the downward slope of the peak, oil and gas prices will fluctuate, but the overall increase will be inexorable. So will the cost of everything produced in an energy-intensive global economy that is predicated on oil's abundance and easy affordability. Not only will heating, cooling, and personal transportation become steadily more expensive; so will our food, produced by industrial agriculture with petroleum-fueled equipment, petroleum-based pesticides, and synthetic fertilizers and then shipped hundreds or thousands of miles to its ultimate consumers by truck, rail, air, or barge. Oil is also the raw material of much plastics manufacture and its rising price will be added to the prices of all the production processes and goods that utilize it. Even if there should be a massive outburst of sanity and hostilities over access to dwindling oil resources don't blow up civilization, Peak Oil's potential to precipitate a worldwide economic collapse still remains fearsome.

Of late, periodicals from *The Economist* ("Oil in Troubled Waters," April 28, 2005) to the goony supermarket tabloid, *The Weekly World News* ("No More Oil! World Supply Will Be Gone in 6 Months!", August 22, 2005) to *E* ("Over a Barrel: Is This the End for OIL?", January/February 2006) have devoted cover stories to the oil situation. Even the CEO of Chevron has informed us, in full-page magazine ads: "It took us 125 years to use the first trillion barrels of oil. We'll use the next trillion in 30."; this followed with an orotund call to "scientists, educators, politicians and policy-makers, environmentalists, leaders of industry…to be part of reshaping the next era of energy."

National Geographic's August, 2005 cover story "After Oil: Powering the Future," examined that next era, emphasizing the development and use of fairly well advanced technologies for solar- and wind-generation of electricity. Electricity, much of which is presently generated by burning natural gas (in preference to the far more abundant and dirtier-burning coal), is critical to industrial and high-tech societies, and a scarcity of electrical power will have different consequences from, albeit equally serious ones, as a drought of liquid fuels. Some prognosticators, notes the *National Geographic,* pin hopes on hydrogen (which is not a fuel, but an energy storage medium), because there are clean ways to generate electricity to crack H_2O, and many believe that hydrogen might function as an oil substitute.

To illustrate the possibilities and problems of a switch to hydrogen, the *National Geographic* presents two entrancing artist's renditions of possible hydrogen futures. In both scenarios, hydrogen would fuel individual automobiles. The first, wherein natural gas or coal would be used to produce hydrogen, looks ugly and stormy. In this coal-burning CO_2-poisoned future, the cars travel bumper-to-bumper on eight-lane freeways. The other scenario, a "zero-emissions technology" hydrogen-powered future, wherein wind, nuclear plants and solar electricity crack water to create hydrogen, looks airy, silvery and agleam. Motoring commuters stream in and out of an opalescent Manhattan that is abristle with skyscrapers taller than any there now.

Apart from the fact that a high-tech megalopolis might not be everyone's ideal, something's missing from this vision. The catch to "supply side" solutions like renewables, hydrogen, or mini-nukes, which technological optimists imagine will forestall the energy crunch, is EROEI: "energy returned on energy invested." How much nonrenewable fuel must be expended in the manufacture of these new gadgets and facilities? After all, the people beavering away on fuel cells will still have to drive to their laboratories on roads made of asphalt (another petroleum product). Our current manufacturing infrastructure, which would be needed to produce such technological fixes, runs on oil. Further, grandiose top-down solutions to an endless demand for cheap energy concentrate investment and, in the case of technologies like nuclear power plants, entail so much security that they rationalize authoritarian social control: the hypothetical technofix would make redundant those billions of human beings for whose benefit the technofix is presumably being initiated, but who don't happen to be experts. Regarding any crusade to switch business as usual over to hydrogen and renewables, Richard Heinberg cautions, "If it can't be done without fossil fuels, it can't be done."

Even if it could, an energy source to power civilization's continued growth and exploitation of natural and human communities might, by further postponing our moral and practical reckoning with other ecological limits, produce an abiotic nightmare of a world, something that would make the first *National Geographic* scenario appear practically sylvan. Even if technology should come up with a substitute for oil, it would have to find substitutes for soil, coral reefs, ocean fauna, forests, water, and indigenous wisdom, all of which are presently being ravaged by global economic activity, underwritten by cheap energy. The demand-side solutions, though, all involve sacrifice and social change.

III. The Price of Admission

My California friend Sarah calls the place where I live "America Land."

Fifteen years ago, upon my return from weeks of travel in India with its throngs of famished beggars, rag pickers, vendors, lean pedicab wallahs, touts, and sleek middle-class pedestrians, America Land, with near-universal auto ownership, free fresh water, plentiful and varied food, and open space left to its own green self-will seemed like a hallucination. Sooner or later, this contrast foretold, an awful lot of us bourgeois naïfs would be awakening from the fantasy of a life without hard physical work.

Did I, given my knowledge and convictions, move then to an ecovillage, start a community garden, implement a two-year food storage plan, or learn to use hand tools and knit my own socks? No. I kept on being a lady writer out here in the woodburb. It's my custom; with minor alterations, it's the ethos and sensibility with which I grew up. I'm living proof that information and experience don't trump inertia and denial. How, then, is it any cause for outrage that titanic outfits like the oil, automobile, real estate, and construction industries, along with research universities, follow civilization's inertia?

This coming cultural-evolutionary correction has been in the making for millennia. Stresses on the tectonic boundaries between subsistence and centralization have been building since cities began. It's not simply that wicked Americans or westerners are prisoners to energy consumption. It's that our species, like all species, is bent on increase. We're invested in being alive, one could say: to reproducing, to inventing, to seeking comfort and abundance. We have more powerful means of exceeding our limits than most species, and we can't go cold turkey on being *Homo sapiens*.

For years, archaeologist Joseph Tainter's *The Collapse of Complex Societies* has informed my thinking about the inherent fragility of civilization, a fragility which predates the fossil fuel era by thousands of years. It turns out that, over the broad sweep of history, civilizational collapse (meaning reversion to simpler forms of social organization) is fairly normal. The centralization of power, the expansion of commerce, and the growth of cities—all implicit in civilization—depend on the production of commodities and storable grain surpluses to sustain urbanite specialists like clerks, clerics, composers and kings. Once human beings gathered in fixed settlements, the majority of most people's time was spent growing food. Only an elite managed to spare itself this labor and invent belief systems to entrench its privilege.

Half the people on Earth now live in cities. In addition to a personal individualism unavailable to the average villager

("City air makes one free" was a medieval saying), cities anciently have been where libraries, orchestras, temples, museums, cathedrals, intelligentsia, bohemias, universities and markets of all kinds—many of the fairest flowers of civilization—flourish. Today cities, with their surrounding slums and residential belts, blanket the arable land that once used to feed their populations. They are severely dependent on extenuated, fossil-fueled lines of supply. Although the present industrial episode has masked this, to grow food is about the most important thing that humans do: the sine qua non. By freeing a goodly portion of humanity from the constraints of the annual solar energy budget (such as the need to labor to grow all the food, fiber, and fuel as well as pasturage for draft animals), the expenditure of ancient sunlight stored in the form of fossil fuels increased agricultural yields, helped depopulate the countryside, and promoted a spike in human numbers. Since I graduated from high school in 1965, the Earth's human population has doubled, from 3 to 6 billion. Another 3 billion fellow passengers are expected by 2050.

As civilizations grow, they must not only extend their imperial reach but also spend more time and taxes to administer and police the ever-more-distant hinterlands upon which they depend. Civilization has proved to be a design for, among other things, exhausting the carrying capacity of its surround. Roughly what Joseph Tainter says in his book is that civilizations always hit a point of diminishing returns. Subjugating the hinterland becomes too costly and, over decades or centuries, things fall apart. In landscapes afar, the cost of oil in blood, nature, and human suffering mounts obscenely. Contention over oil plagues Latin America and Africa as fiercely as it does the Middle East. "The industrial world's addiction to oil is laying waste to Africa. We gas up our SUVs with these people's lives," wrote David Morse of the three to four hundred Darfurians dying every day in the genocidal conflict over oil revenues in the Sudan.

The resource-intensive, fossil fuel-powered American way of life enables the upper tier of North Americans to live like pashas of yore, but without the same face-to-face inhumanity. Richard Heinberg's back-of-the-envelope calculation is that burning oil provides each of us with the equivalent of 300 human slaves' labor. To say, "The American way of life is not negotiable," as an earlier George Bush did, was candid, if void of humanitarian concern or foresight. However difficult, using less oil instead of waging war to control the supply would declare a hopeful preference for community over colonialism and for sun, soil, and muscle as power sources in preference to centralized technologies and expert rule.

Whether or not it ultimately topples the empire, the concept of Peak Oil begs the question of how we are to live. As actors or subjects? Exploiters or kin? We must acknowledge that the vast majority of us are on the same foundering ship. A few yachts will sail away, leaving fate to fix the rest of us up with some interesting flotsam-fellows. One may still be able to chitchat within a virtual community intermittently, as electricity is available, but one's geographic neighbors will be the persons to deal with daily. Our various ideas about what is reasonable, desirable and possible will have to be reconciled on the ground.

My widowed friend Naomi spent her married life as the spouse of a General Motors chief engineer. Naomi's respect for Cadillac, GM, and technology generally is exalted, and our conversations are mutually challenging. After one of these last summer, I lent her the DVD of *The End of Suburbia: Oil Depletion and the Collapse of the American Dream*, an alarming Canadian documentary about the folly of basing a whole way of life on cheap oil. Naomi's comment was one I could only agree with: it didn't leave any room for hope. Scaring the pants off people seldom promotes creative enthusiasm. *Au contraire.* Following my discovery of Peak Oil and a winter spent cringing under the covers at the thought of a Mad Max style—*dénouement* of America Land, I felt put upon. Instead of riding my bicycle to a little inland lake nearby, I spent the summer comforting myself with swims at a choice Lake Michigan beach a twelve-mile drive away. "As long as you can afford the gasoline," whispered my id, "why not?"

One day at lunch, after I had rapped out another survivalist-doomsday scenario of exorbitant diesel costs choking off the truck traffic that makes our supermarkets so reliable and the individual automobile going the way of the top hat, Naomi asked what was going to employ people in the post peak, post-collapse America.

"Growing food," I replied.

Having cast an eye over my pitiful garden, she said, "Well then, you better get some black dirt and cow manure in here and start cutting down some of these trees."

IV. CASHING THE REALITY CHECKS

A while back I began to read nineteenth-century English and American novels for deep pleasure. Among other riches, these furnish my mind with images of a life that never knew the diversions and physical comforts that seem so normal to us today: elevators, indoor plumbing, central heating, sewage treatment, snowplows, fork lifts, washer-dryer combinations, family cars, chain saws, wood-splitters vermin-free beds, electric light, gas cookstoves, and frequent changes of underclothes. We have it on good literary authority that such lives have been not only humanly possible but rich in meaning.

Though I have a general idea of a way of life I'd wish for (place-specific, cooperative, diverse, healthy, inventive, capable, earth-healing, and Arcadian) on the other side of Hubbert's peak, the way to go that distance isn't self-evident. To judge from the excitement about Peak Oil on the Internet, a good many people are jumping in feet first to make the road by walking. Apparently, they're past shock, denial, inertia, and fear and have moved on to acceptance. Most important, they are no longer waiting for MIT or the Feds to rescue them. It does seem almost possible that in the twenty-second century, human-scale cities and towns will be the capitals of self-reliant watersheds.

That strident movie *The End of Suburbia* is a bolt of information that's stimulated a flock of community initiatives. The Northern California lumber town Willits responded to the prospect of Peak Oil in an exemplary, but not unique, way. In October 2004, biologist Jason Bradford, a relative newcomer to Willits, sponsored some *End of Suburbia* screenings. Rather than paralyzing the numerous townsfolk who attended, it mobilized quite a few, including some local government officials. Built on a substrate of "New Settler" culture (that described in an eponymous Northern California periodical), the Willits Economic Localization Project (WELL) got underway.

"Few people have control over basic necessities in their lives," wrote WELL's founders. "Multinational corporations don't care about our town. Distant government agencies don't know what we need; being dependent on nonrenewable resources from politically unstable parts of the world threatens our security. As responsible adults, we'd rather face reality and do what is necessary to avert catastrophic climate change, mass extinction of species and the loss of democratic freedoms."

Using all the data they could get their hands on, WELL conducted area energy and food supply inventories, calculated the gas money that relocalization could husband for the town's economy, did the math on the acreage required to keep the town fed, and enlisted the interest of the local American Legion post. They've held a disaster preparedness forum, developed a Biointensive Garden Plan and evidently inspired the digging of community gardens all over the place. Willits is doing, rather than simply talking about, grassroots, participatory strategic planning.

Be it noted that many, many meetings are required to keep something like this moving. Of course, if those meetings include congenial, inspiring people, use consensual rather than parliamentary process, and help participants to arise from the slough of despond, then their frequency isn't a drawback. As part of such a group at home in Michigan, the Traverse Area Community Currency Corporation (TACCC), I'll vouch for that. Our endeavor isn't as comprehensive as the WELL, but it was founded in light of, and has direct relevance to, the economic shakedown that the peak may provoke.

TACCC was started a few years ago to develop an alternative currency for the Traverse City area. Its board is a pool of talented, lively, and effectual people like my friend Sharon, who helped start an (alas, defunct) car-sharing club in Traverse City and mastered an awful lot of information about the oil-based transportation system in the process (she looks forward to the price of oil rising enough finally to pry us out of out cars and back into conviviality); Bob, a former science teacher, now an innkeeper and the president of a local peace and ecology center that, among other good works, annually hosts the Great Lakes Bioneers Conference; Liz, an MBA, past president of the local natural foods cooperative, and the operations director of a winery; Jody, who lived a while in Chiapas where she cofounded a 100%-fair-trade coffee roasting company, whose profits build clean water systems in the villages where their coffee is grown; Brad who, when he isn't researching local currency or fine-tuning our web site, is a community organizer, musician, carpenter, and designer. He and Amanda, his wife, work to develop a nonprofit that will link art and sustainability. And there's Bill, the veep and landlord: a professional facilitator who, in his day has run a listener-sponsored radio station and a gourmet bed and breakfast on an island in the middle of Lake Michigan.

Notwithstanding their individual accomplishments, the TACCC board is, in my estimation, not a particularly exceptional bunch. Occasionally I lecture on ecology and social change here in the Midland and so visit lots of small and medium-sized cities with thriving green (human) populations. On this basis, I would be surprised not to find any community of a certain size without a similar core group of people whose skills and civic ambition are up to helping navigate the Peak Oil transition. We board members of TACCC have varied reasons for participating, differing visions of the future, and varying senses of the urgency of our project. What unites us, however, is a belief in the importance of economic relocalization, which means reinstating local production for local needs. The Peak

Oil angle to it was one of my motivations for volunteering. By working on local currency, I'm planing a plank for the regional lifeboat. After studying our situation for some years, I had come to believe that, because the value of the U.S. dollar is partly bound up with the commerce in oil and largely dependent on continuing economic growth and the benevolence of our creditors and because economic decline is apt to be among the presenting symptoms of declining oil and gas supplies and rising costs, the official national currency could lose much of its value. Of course, national and global economies are monumentally complex systems, hinging on perceptions almost more than physical reality. Still, having a functional local currency to fall back on could well supply a workable means for the exchange of goods and services. In fact, this is not the first local currency in local history. Traverse City had one in the Great Depression.

The introduction of Bay Bucks is my current non-rhetorical work to abet a deliberate, not a precipitate, relocalization of the globalized economy. What's more, I am happy to report that I recently overcame my reclusiveness and invited the new neighbors over to dinner. Although I still don't garden, I've cultivated a happy relationship with a nearby Community Supported Agriculture (CSA) farm and am swamped now with the handsomest produce I've ever seen.

Even if we should all walk away from last-ditch efforts to find substitutes for oil and gas that will feed the growth monster and so protract the American Way of Life and instead apply ourselves to the challenges of our local commonwealths, we'll still face chaos and difficulty. Post-peak planetary civilization will probably look like a mosaic of ruin, reclamation, and resurgence across the planet. The next few hundred years will be a testing time, especially for the civilized portion of our species and for those of us still residing in America Land. Our communities must find their right relation to an altered planetary ecology. If the lifestyle to date has been like living in a hotel, the lifestyle to come will be a camping trip, or maybe even a lifetime Outward Bound course.

Bioregionalist Peter Berg speaks of "taking on the big questions from a place small enough to yield practical results," and I swear by his wisdom. Hope is like lichen, growing close to the ground, making do with rocks and water, building soil again. Real live geographic community is the only kind of place to dig in to face the challenges that declining energy availability, economic depression, climate change and public health crises portend. Only in place will we learn anew, in myriad ways, to carry on.

In the meantime, whenever I take a hot shower in water clean enough to drink, eat chilled cherries, switch on an electric fan or light; or drive my car to a distant beach or crawl into a warm clean bed, I try to make the most of it. As long as I still have the leeway not to grow my own food or look for an opening as a serf, I'm enjoying my desk job. But I have no doubt that the party's over, and that may be a good thing.

INSIDE AGITATION

LEARNING TO LIVE WITH AMBIGUITY

Any credentials that I may have to pontificate began to accrue when I gave the 1969 commencement address at Mills College, my alma mater. There, before the graduating class of seventy-five young women, I explained why I thought "The Future Is a Cruel Hoax." During my last year of college, I had become quite concerned with environmental problems, feared that our rosy future was imperiled by these, and felt that it had become necessary to sound a general alarm.

At that point I focused on the problem of overpopulation for two reasons: one, the brilliant persuasion of Paul R. Ehrlich; and two, my naive yearning for explanatory principles. Ehrlich's *The Population Bomb* gave my natural-born pessimism plenty to resonate with. In the speech, I gloomily predicted widespread famines and plagues, saying that we were breeding ourselves out of existence. I predicted our extinction within less than ten years and announced that, as a result of all this mess, I was "terribly saddened that the most humane thing for me to do is to have no children at all." For reasons both altruistic and selfish, I honored that commitment till age mooted it.

Saying all those dire things and opting out of parenthood so early made me instantly, if momentarily, famous and catapulted me into the ecology movement, in which I have found myself occupied ever since.

The ensuing years taught me that explanatory principles can be dangerous and that they don't make ecology like they used to. In the late sixties, it was possible to succeed as an environmentalist on the strength (and flash) of one's pessimism alone. Pessimists are never disappointed, I reasoned. Ecocatastrophe was news then, and it lodged a feeling of hopelessness inside me. I believed what I was saying; it wasn't empty rhetoric. I felt hoaxed, cheated of a nice safe future. Yet for me to have believed that there ever was such a thing as a nice safe future should tell you something of my youthful ignorance of history.

For history tells us that empires come and go. Environmental exploitation is nothing new; and war is the rule, not the exception. If the enormity of the consequences of these typical human activities has increased, so has the scope of our attempts to change. I can accept that now, but in '69, Paul Ehrlich's quip that "nature bats last" seemed about right, and I figured that we, the visiting team, were about to lose in disgrace. Fine with me. Subconsciously (or not so), I preferred to see the apocalypse than to work on through the task of living in a less than perfect world.

I was an idealist. I regarded humanity in aggregate and blamed it so. Therefore, in addition to being then and now a monster of a problem, overpopulation dovetailed with my psychology.

Not long after my graduation, an appointment to the Board of Directors of the Planned Parenthood Federation of America enabled me to pursue and broaden my concern more or less actively for six years. My colleagues on the board were an impressive group of men and women, who possessed a variety of expertise and a lot of civic-mindedness. Our stated goal was to make every child a wanted child and, discreetly, to educate the prospective parents of the world to want fewer of them. At the time I believed that birth control was an unmitigated boon and population control a necessary and not unmitigated evil. And, because I had no doubts about the high quality of my motivations, I worked hard at institutionalizing my own brand of good and evil. It seemed like a simple enough proposition. At that time, my tolerance for ambiguity was nil. I didn't understand that it is impossible to do only the thing you want to.

Working at the national and sometimes international level, the Planned Parenthood Federation of America Board looked, of necessity, at the big issues: the population explosion, the revolution in sexuality, the unending struggle for reproductive choice. We hoped to serve the millions of women and men who needed contraception and to secure the right to abortion for all. Because, at the time, they seemed to be the most effective and efficient means of birth control, we promoted the use of pills and IUDs. These appeared to be panaceas for the problem of unwanted pregnancy. But, as Stewart Brand astutely put it, "Panaceas are always poison."

A twenty-year-old skier, a birth control pill user, breaks her leg; has it set in a cast. A blood clot develops in her immobilized calf, she can't feel its warning pain, and it travels to her brain, causing a stroke and paralyzing her.

A thirty-five-year-old lawyer, having worn an IUD for seven years with little apparent ill effect, having established her career and decided to start a family, begins to try to get pregnant. But she can't. She learns that her fallopian tubes have been scarred as a result of a long-simmering case of pelvic inflammatory disease, aggravated by the device.

Such things really do happen. Not a lot, but once is enough if it happens to you. As advocate then, and dinner-table arguer now for free access to birth control, I feel some responsibility for whatever adverse effects birth control efforts have and for whatever progress women have been able to make as a result of being free to choose whether to become mothers. My point is that when you get to messing around with social change, you have to stand ready to acknowledge your victims as well as your victories. Experience has convinced me that it is impossible to know with certainty that your cause is good.

Even more confounding, I find that, close up, villains appear less villainous; their evil deeds less incomprehensible.

I've got a friend whom some might call a war criminal. He worked for the CIA in Cambodia during the war in Vietnam. His activities were a small contribution to the destruction of a people, the final outcome of the Nixon-Kissinger sideshow in neutral territory. My friend married a Cambodian woman, wore a sarong, lived in a hootch, and, when he had to, killed a Vietnamese officer with his bare hands after an endless, harrowing, cat-and-mouse chase around a village well.

My friend is a thoughtful person. He believed in what he was doing over there. He was incapable of thinking it all the way through, just as I was. After a few years of nightmares that had him drenching whole mattresses with sweat and a few more years of trying to ignore his deeds, he's found a way to atone. He's working with Cambodian refugees in San Francisco: just a family that he's able to help in his spare time.

He speaks their language fluently. "I spend an hour or two trying to translate Cambodian into English. When they say 'hello' they're really saying 'Good to see you and how's your health?' They say things like 'Go in peace' instead of 'Good afternoon'." He knows and admires their culture; he wants to help them to communicate their way of being more clearly, and he is one of the few do-gooders around with a concrete understanding of the difficulties of their situation here. For instance, a lot of the Cambodian-speaking welfare workers who help the few refugees who made it out are Vietnamese, who have a long-standing and reciprocated enmity with Cambodians.

My friend knows he can't patronize these people. "I want to be the family's friend. I think they need to be treated with dignity. I don't want to promise them stuff that I can't change. I've got to get down off the white charger. I know it's gotta be neat for somebody to watch it, but I'm not doing it for an audience." He is helping them because he needs to. Selfish reasons. He knows exactly why they got here and what they left behind. And he seems to know the difference between saving his own soul and assuaging his guilt. That difference lies in forward motion.

This friend has a lot of terrible insights to share, and I like him for it very much. When we get to talking at the level of mortal consequences, I get terrible insights too. They undermine my certainty and therefore make me less innocently dangerous than I might be. Like me, he was involved in what he thought was a good big cause. Like me, he played a small part in a big change and now finds that doing good is in the particulars.

It was in talking with him about his taking of human life that I realized that abortion may be woman's version of killing in combat; this analogy makes sense to me at least. And on my own battlefront, I value the right-to-life movement for calling attention to the spiritual consequence of abortion and for asserting the marvelousness of conception and the sacredness of every single germ of life.

That doesn't mean I wouldn't have an abortion if I needed to. It doesn't mean that I agree with the policies the right-to-life movement seeks to impose on this country. It does mean that I would appreciate the gravity of the choice. I want to live in a society where people do make grave choices, exercising individual conscience. Dictating those choices and proscribing the exercise of conscience thereby causes it to atrophy. Such mortal choices make us confront a reality which is forever more complex and less than ideal.

There was a time when my family-plannish idealism caused me to despise any right-to-lifer. A few years later, however, my attempt to embrace reality found me and a few billion others admiring Mother Teresa, a right-to-lifer extraordinaire. When she accepted the Nobel Peace Prize, she said, "To me, the nations with legalized abortion are the poorest nations.... The greatest destroyer of peace today is the crime against the innocent unborn child." On that, she and I would have to agree to disagree. One of her lay assistants was quoted in a *New York Times Magazine* story as saying, "Mother has her beliefs and she holds them sincerely and anyway, her specialty is love, not moral philosophy."

For saints, we make allowances. Her particular acts of mercy, her charity, can be a beacon to us all. To bathe and feed and dignify the dying of Calcutta's poor is as real as it gets. The *Times's* Michael Kaufman put it thus: "It is part of Mother Teresa's very concrete philosophy that the world needs the poor as a reservoir of love." She says, "Our poor people are great people, a very lovable people. They don't need our pity and sympathy. They need our understanding love and they need our respect. We need to tell the poor that they are somebody to us, that they, too, have been created by the same loving hand of God, to love and be loved."

Objectively viewed, Mother Teresa looked like Sisyphus. The Missionaries of Charity may never make statistically significant inroads on the problem of poverty ("We ourselves feel that what we are doing is just a drop in the ocean," said Mother T.) but the kindnesses attributed to her and the religious and lay members of the order that she founded appeared to be drops of grace sparkling over an ocean of woe.

Decades after announcing the apocalypse, I realize that I probably won't have to face the End of the World in my lifetime. I see that there's work to do and have learned that doing unalloyed, or sufficient, good is another impossibility, unless that good is so specific that it may seem insignificant to the faithless and invisible to demographers. Mother Teresa put it correctly: "I do not agree with the big way of doing things. To us what matters is an individual. To get to love the person we must come in close contact with him. If we wait till we get the numbers, then we will be lost in the numbers."

I still worry about population. I wanted to *do* something to change it. Perhaps I did, but I also inadvertently abetted some bad things, to the extent that my enthusiastic embrace of global solutions to poignant problems that play out in the lives of billions of individuals led to more, unanticipated problems. To the extent that my little contribution enabled some women to become moral actors and make serious choices in their lives, I'm proud.

I'm still more willing to talk about what's wrong with the world and to daydream useful ideas for others to carry out than to change my own life accordingly. For years, I was daily entertained by the uncertainty and the gentle mocking irony of my lifestyle. Uncertain: I wondered time and again whether I should have a child someday, take a little chance on the future, and admit some of the life-enhancing ("growth-demanding," one friend called it) chaos that babies bring. Ironic: I loved city life, being part of the imperium, reading in bed by electric light, and driving my car, all in the service of the ecological movement. Even now that I live in the woodburbs, I eschew the hands-on toil it takes to restore the Earth.

That makes me an ordinary human, and it is beginning to give me a little patience with the world's slowness to reform. Gandhis, Dorothy Days, and Martin Luther King, Jrs. only come along once in a generation or so. On close inspection, they aren't perfect, either. But their conscience, their courage, and their adamant nonviolence are ideals that any of us can try to approximate, each in our ways.

Hope is the basic requirement for that kind of work. Folks don't go to jail over A-bombs, civil rights, or the wrongness of war because they're expecting to lose. They can't lose any more than Mother Teresa can fail. They may not achieve a perfect good, but they will have labored in service of a higher purpose, and it isn't over 'til it's over.

These days my advice is to learn to live with ambiguity. I could urge unambiguous nonviolence, but in the lives of some I've known, and someday perhaps in mine, push does come to shove. I know that my unambiguous idealism has done violence to reality. One could spend several lifetimes in the attempt to expunge every last trace of invidious judgment from one's idealism.

It's ineluctable: irony is the human condition, imperfection is a leading characteristic of heroes, and you can never do just one thing.

SHUTTLING
BETWEEN WRATH
AND COMPASSION

(1989)

Some years ago, the women of Birthways Midwifery Service asked me to serve as the mistress of ceremonies at their Mother's Day Peace Pole planting celebration. The invitation came more on account of my verbal facility than for any particular knowledge or dedication of mine to either peace or motherhood. I don't do peace (just as some house-cleaners don't do windows), and motherhood was definitely not my calling. Nevertheless, I was honored to be thought of in connection with those issues, and by individuals for whom I have great respect. Naturally, I accepted.

Besides, it amounted to being asked an interesting question—what is peace, anyway?—and a question is ever an opportunity to learn.

I had to ask myself, "Why isn't peace my issue?" Perhaps it's because peace is already addressed with an enthusiasm that radical ecology, which *is* my issue, doesn't quite seem to engender. Peace, if it may be defined simple-mindedly as the absence of war, is in the obvious self-interest of just about all human beings except the czars of the military-industrial complex. (And radical ecology, while in the interest of the human species, could inconvenience a lot of human individuals.) There are apparently a few human beings whose greed is satisfied by manufacturing engines of war and people whose personality structures may be fulfilled by regimentation and the giving or following of orders—however absurd, destructive, or suicidal—but, although some people will admit a preference for war if the choice is between that and a threat to an abstraction like democracy, communism, freedom, order, or national security, virtually no one, even an arms maker, will admit to desiring war.

So peace, I thought to myself, is a motherhood issue. Very popular these days. But what is it? And is it a natural part of the human behavioral repertoire? We talk about peace of mind and we experience it every once in a while, but one school of wisdom proposes that peace on Earth will stem from a flourishing of peaceful minds. "Let there be peace on Earth and let it begin with me," goes the song, with its ring of truth. And the *Book of Common Prayer* has the priest give this blessing: "May the peace that passeth all understanding be among you and remain with you always." That peace should pass all understanding suggests that it is a rarity and a mystery, something inward, perhaps even irrational.

The image of a baby at the breast, of a mother and child reunited as one body by the nature of nurture, is transcendently peaceful, hinting at the fundamental justice that is prerequisite to social peace, the justice of support and sustenance provided as a simple birthright. So at both the symbolic and literal levels, it is very apt that a Mother's Day celebration should refer to peace. The Birthways invitation credited the abolitionist and social reformer Julia Ward Howe with having instituted the holiday. Before the greeting card and candy mongers seized upon it as a sentimental occasion just perfect for a marketing orgy, it was called Mother's Peace Day and was launched with serious intent. Julia Ward Howe's 1870 Mother's Day Proclamation read in part:

> Arise, then women of this day! Arise all women who have hearts, whether your baptism be that of water or of tears. Say firmly: "We will not have great questions decided by irrelevant agencies, our husbands shall not come to us reeking with carnage, for caresses and applause. Our sons shall not be taken from us to unlearn all that we have been able to teach them of charity, mercy and patience. We women of one country will be too tender of those of another country to allow our sons to be trained to injure theirs." From the bosom of the devastated Earth a voice goes up with our own. It says, "Disarm! Disarm! The sword of murder is not the balance of justice." Blood does not wipe out dishonor nor violence indicate possession. As men have often forsaken the plow and the anvil at the summons of war, let women now leave all that may be left of home for a great and earnest day of counsel. Let them meet first as women, to bewail and commemorate the dead. Let them solemnly take counsel with each other as to the means whereby the great human family can live in peace, each bearing after his own time the sacred impress, not of Caesar, but of God.

Howe's tone is hardly peaceful. Rather, there is a sense of outrage at the very existence of war (this was the lady, remember, who wrote about trampling out the vintage where the grapes of wrath are stored). She is, among other things, calling for a withdrawal from complicity with murder. She makes common cause with other givers and nurturers of life. She will not concede any legitimacy to war, and the genius is her indignation at the brute stupidity of it.

In a tiny little book called *Your Reason and Blake's System*, Allen Ginsberg wrote:

> The prophetic books are actually reflections of William Blake's personal conflicts of the time. In *Jerusalem*, there is a theme which is useful now: the argument between political anger—say over the nuclear bomb—and a sense of compassion and mercy; and a realization that the World doesn't matter, or that if it does matter, there's no way of approaching it with anger. Blake was struggling with some of the same emotions we struggle with, which, I assume are more or less common, for his revolutionary times—post French Revolution—and the destruction of idealism, radical disillusionment.

> There are similar revolutionary conditions now as in Blake's time, similar social and emotional problems. Blake's books are useful now as explorations of the same problems we have, somewhat related to the revolutionary fervor of the sixties in America and a subsequent so-called "disillusionment." So, actually, Blake is up to date in the psychology of wrath versus pity, compassion versus anger, that runs through all of his work and is visible for our own time as well as his.

The tension that Ginsberg identifies—between wrath and pity, between compassion and anger—is very familiar to me, although it is provoked less by the thought of the bomb than by the sight of tree stumps. The war that I am really concerned about abolishing and which preceded and has ever since been implicated in all other war, is the war against the Earth:

World War III, conservationist Raymond Dasmann called it. It seems to me that to end this war on the biosphere and to secure evolutionary justice, not only for the human species, but for everything that lives ("everything that lives is holy," wrote Blake) will require of us choices and changes that will make the attainment of social peace seem easy by comparison. Because of the evidence all around of our endless, witless war against everything that is not human or to human taste, I find myself shuttling constantly between wrath and compassion. I am wrathful that we seem at times to be so stupid about our fundamental inseparability from the biosphere—the fate of the earth is the fate of humanity—and I ache with compassion for the wounded hubris that makes us that way.

In the short term, anyway, ire is a strangely gratifying emotion. It is often hard to contain wrath, and it is even more difficult to bear the feeling of compassion: it takes sustained energy, and it softens up the borders. To identify with the Other or with the All is a daring proposition. It entails the loss of the separate self. We have selves for some reason, and time spent without them is naked eternity.

Consciousness, however, seems often to lead in that direction. I am honored to participate in anything that midwives do because of their closeness to birth—at once the most commonplace and the most marvelous experience of the unity and separation of selves—and because of the tough compassion that midwives must practice. Conscious childbearing is part of restoring to sacredness love, sex, gestation, birth and nurture, aging, and dying. Witnessing these unbinds the heart and startles the mind. Among the greatest gifts anyone has ever given me was the privilege of attending the birth of her child. The friend who bestowed this gave birth to her boy with grace and serenity and real hard labor. Being present with her and her husband and her first son at that birth, I glimpsed major Earth-magic. First there's just the mother and the father and the onlookers and then, a second later, another whole person is present—slippery and squally, to be sure—but another being has swum into life, just like that. The beginning of life and its ending are our paramount mysteries.

As planting time gets into full swing, many of us are plunging enthusiastically into the pleasure of gardening, one of the greatest of the arts of peace. In the garden or in the woods, we can witness Gaia, the Mother of all, busily engaged in sex, gestation, birth, and nurture, aging and dying. If we pay attention to all of this, the Earth can teach us what is involved in establishing peace and creating beauty; what is involved in achieving balance, abundance, and continuity. The information we need, the wisdom to accomplish these things and to cultivate these qualities in our own beings, is everywhere in nature if we will but read the lesson. There we can even learn about conflict and competition and the taking and ending of life, all of which must and may go on, and which only becomes devastation when one species pits itself against all others, and against itself, and the balance is disrupted.

"All flesh is grass," proclaimed the prophet Isaiah, "and all the goodliness thereof is as the flower of the field." It is the order of nature that flesh, like grass, is eaten. Isaiah was no vegetarian and neither is the hawk owl that swoops out of the spruce hungrily to reduce a grouse to fluff, shins and tail feathers. Such is not war. It is part of a larger, vital peace.

THE MATRIX OF
SOLITUDE

(2002)

No two solitudes are alike, and any solitude spent in nature's company will brim with incident and mystery. I spent the first September day of 2002 enjoying such a richness, looking around and gazing within. It was a warm late summer Sabbath in the north woods countryside, pleasantly free of the workaday noise of commerce. From Monday through Saturday a nearby highway and busy landfill raise a mighty mechanical din. The Sunday quiet left room for the gentler sounds of cicadas, crickets, rustling leaves, and distant lawn mowers.

I call myself an ecological activist. Over the past thirty years, in both the city and the country, my activism has consisted of wordcraft; of advocacy and agitation; of books, articles and oratory in service of the wild. For as the master said, "In wildness is the preservation of the world." Whether the wildness happens to be that of an old-growth Lake States forest, of the monarch butterfly migration, of individual conscience, or of social invention and creative dissent, I seek its flourishing to preserve a world for me.

Like my hero, Henry David Thoreau, I live outside of town by choice. Thoreau was wary of enterprises that require new clothes. By 1984, I was wary of rents that required holding jobs in which I lacked conviction, so I left the city of San Francisco for the country of northwest lower Michigan. Also like Thoreau, I was willing to simplify my life in order to obtain the time and space in which a book could be got together.

My woodland home fronts me with a more than human community whose parts and wholes are all alive and beautiful, comprised in Nature's self-willed integrity. My encounter, as a solitary human being, with this community supplies me renewably and reliably with my means and ends.

"When I detect a beauty in any of the recesses of nature," Thoreau wrote, "I am reminded by the serene and retired spirit in which it requires to be contemplated, of the inexpressible privacy of a life."

On that recent Sunday by myself I sat on my back porch sipping my morning tea. Noticing that the hummingbird feeder that hangs from a branch in a young maple tree was empty, I got up and filled it. While I was about that little task, a caterpillar, a maple spanworm, must have climbed my pajama leg. This moth larva's smooth skin was exactly the rouge-grey color of young maple bark. Near its posterior a protuberance even mimicked the look of a joint in a twig. When on my return to my tuffet I first noticed the perfectly camouflaged critter, it looked like a little twig. Yet the twig startlingly became an inchworm compassing brisk counterclockwise circles on the striped fabric of my pjs After I had contemplated it awhile, I tried to brush it to the ground. The tenacious spanworm reared up, struck a stiff twig pose and clung to the pajamas. At length I pried it loose with a stem of dried grass. The caterpillar remained rigidly in twig character as I transported it back to the maple tree.

Although there is much that is exemplary, good, and hopeful in our species, I look outside my kind for wholeness. I would die inside without the ever varying, always astonishing presence of nature. It is not merely a storehouse of metaphor. Life's examples present truths that baffle words.

When I am by myself under the open sky, my thoughts are free to roam. My being and my vicinity seem to be at one and imbued with joy. It occurs to me that solitude is a likely place to know peace and freedom, those two great desiderata of activists throughout the ages. I work to translate my experience into prose that may be as meaningful and useful as Thoreau's has been to me and to innumerable lovers of nature and simplicity. Thoreau, as a great exponent of both solitude and civil disobedience, is both my idol and a source of reassurance that although I may be solitary, I am not alone in my temperament.

However collectively they may work in pursuit of some vision of a better world, activists are usually isolates in some respect. For activism implies objection, criticism, judgment and nonconformity quite as much as it implies solidarity with a cause and the affirmation of some objective.

He is thought of as antisocial but Thoreau's solitude was actually quite permeable. He lived a rustic life of food gardening, sauntering, observation, and exposition of nature, all the while maintaining friendships with fellow writers and townspeople and enjoying close relationships with his family. In his house, he wrote, were three chairs: "one for solitude, two for friendship, three for society."

His conscience and integrity, forged in his family, wrought in solitude, and tempered in nature, would change the course of history. "On the Duty of Civil Disobedience," printed in 1849, emerged from Thoreau's refusal to pay his poll tax as a protest against the United States's war on Mexico. Thoreau decided to be jailed rather than pay the back taxes and become complicit in the U.S. aggression that bid, through territorial expansion, to increase the number of slave-holding states in the Union. Although a well-meaning aunt paid his back taxes the next day, Thoreau's essay, upshot of a night in the lockup, would inspire Tolstoy, Gandhi, and Martin Luther King, Jr., a whole modern lineage of nonviolent direct action.

The mass mind is incompatible with the deep questioning and original thinking that may be mulched in solitude. Those who hope to see the wild possibilities inherent in human souls blossom in full diversity had best honor solitude, whether institutional or irreligious. Perhaps the minds of hermits, yogis and yoginis, of nuns and monks in silent cloisters, and of cussed recluses all are wilderness refuges for honor, clarity of vision, and the evolution of ideas—like the duty of civil disobedience—to nudge humanity toward better ways.

Stephanie Mills is an author, lecturer, and longtime bioregionalist. Her books include *In Service of the Wild: Restoring and Reinhabiting Damaged Land* and *Epicurean Simplicity.* Since her emergence in 1969 as an ecological activist, Mills has written prolifically, edited numerous periodicals, participated in countless conferences and served on the boards and advisory committees of dozens of ecologically oriented organizations from the local to national level.

Since 1984 she has lived and worked in northwest lower Michigan. She is presently at work on a biography of Robert Swann.

Acknowledgements

In slightly different form, these writings appeared previously as follows: "The Journey Home" published in *Sierra*, September/October 1997; "Tough Little Beauties" web published by *Your Place*, September 2006; "Reverence for Forests, Reverence for Wood" written for *Trees for Life*, 1999; "Paper Birch" published in *Synapse*, Fall Equinox 1996; "India Journals" published in *Synapse*, Winter Solstice, 1992; "On Fetal Personhood" written for *Whole Earth* no. 51, Summer 1986; "Nulliparity and a Cruel Hoax Revisited" published in *Wild Earth*, vol. 7, no. 4; "Is the Body Obsolete?" written for *Whole Earth* no. 63, Summer 1989; "Spiritual Swimming" published in *New Age*, July 1983, vol. 8, no. 12; "St. Herpes" published in *Whole Earth* no. 48, Fall 1985; "Some Words for the Wild" published in *Resurgence* no. 205, September/October 2001; "Fun While it Lasted" web published by *Energy Bulletin*, Thursday June 22, 2006; "Learning to Live With Ambiguity" published in *CoEvolution Quarterly* no. 28, Winter 1980; "Shuttling Between Wrath and Compassion" published in *Synapse*, Winter Solstice 1989; "The Matrix of Solitude" published in *Ascent*, Winter 2002.

The Ice Cube Press began publishing in 1993 to focus on how
to best live with the natural world and understand how people
can best live together in the community they inhabit. Since this
time, we've been recognized by a number of well-known writers,
including Gary Snyder, Gene Logsdon, Wes Jackson, Annie
Dillard, Kathleen Norris, and Barry Lopez. We've published a
number of well-known authors as well, including Mary Swander,
Jim Heynen, Mary Pipher, Bill Holm, Carol Bly, Marvin Bell,
Ted Kooser, Stephanie Mills, Bill McKibben and Paul Gruchow.
Check out our books at our web site, with booksellers, or
at museum shops, then discover why we are dedicated to
"hearing the other side."

Ice Cube Press
205 N Front Street
North Liberty, Iowa 52317-9302
p 319/626-2055 f 413/451-0223
steve@icecubepress.com
www.icecubepress.com

from high and low, near and far
thanks, hugs, kisses and cheers to
Fenna Marie & Laura Lee

Other Ice Cube Press books of interest include:

River East, River West: Iowa's Natural Borders
1-888160-24-1 , $12.95
Writings by David Hamilton, John Price, Gary Holthaus, Lisa Knopp, and Robert Wolf, with "creekography" by Ethan Hirsh, on the history, folklore, nature and ideas of the two rivers bordering the state of Iowa. As this book explains, the Mississippi and Missouri rivers are much more than the water that flows in them.

Prairie Weather
1-888160-17-9, $10
Iowa is at the crossroads of the elements—just above our heads whirl other-worldly tornadoes, and summers bring bone-drying droughts, while winter brings walls of snow. In our region of four seasons, we can learn much from our weather. Writing and photographs by Jim Heynen, Mary Swander, Deb Marquart, Amy Kolen, Ron Sandvik, Mark Petrick, Ethan Hirsh, Robert Sayre, Thomas Dean, Patrick Irelan, Michael Harker, Scott Cawelti, and a foreword by Denny Frary.

Living With Topsoil: Tending Our Land
1-888160-99-3, $9.95
A full-fledged exploration via Iowa's finest authors into living with our state's world-famous topsoil. New and valuable writing by Mary Swander, Connie Mutel, Michael Carey, Patrick Irelan, Thomas Dean, Larry Stone and Tim Fay, and an introduction by Steve Semken. Jose Ortega y Gassett once wrote, "Tell me where you live and I'll tell you who you are." You'll find out what it means to live in the land of amazing topsoil once you read this book

Find out more about all our titles go to
www.icecubepress.com
or by mail (check/money order) by sending to
The Ice Cube Press
205 North Front Street
North Liberty, Iowa 52317-9302
(shipping $1.50 first book, .25¢ each additional)